THE MISTS OF SPECIAL RELATIVITY

Time, Consciousness and a Deep Illusion in Physics

Stephen Earle Robbins PhD

ISBN: 1492900427
ISBN 13: 9781492900429

Stephen E. Robbins, Ph.D.
Golden Willows Farms
2750 Church Rd
Jackson, WI 53037

SEarleRobbins@yahoo.com
stephenerobbins.com

Table of Contents

Preface

I must warn the reader at once that this book, 1) is only a somewhat modified version of the last chapter – a chapter devoted to the subject of relativity – of my earlier book, *Time and Memory: A Primer on the Scientific Mysticism of Consciousness*, and, 2) contains a very condensed version of the theory of conscious perception laid out in that work, just enough to sketch the considerations that are relevant to relativity herein. Therefore, for anyone who has read the *Time and Memory* work, please be warned not to expect all that much in the way of difference.

My reasons for interest in this subject arose while doing research on my doctoral thesis (1976). The thesis bore a similar title to the above book, namely, "Time and Memory: The Basis for a Semantic-Directed Processor and its Meaning for Education." The thesis was developing Henri Bergson's model of perception and memory, a model embodied in his 1896 work, *Matter and Memory*. I had seen that Bergson's model was in essence holographic – an insight of Bergson fifty years before the discovery of holography by Gabor – and likely the prime reason why its model of perception was considered obscure by Bergson's contemporaries, in fact, for a complete failure of comprehension by subsequent philosophers and commentators on Bergson. It held a radically different use of holographic principles compared to the conception which became popular via Karl Pribram and which, unfortunately, has kept the subject of holography and perception in a straight jacket ever since. The thesis was updating Bergson in modern terms, using the principles of holography and the theory of perception of the great theorist of perception (and mentor to my thesis advisor, Bob Shaw), J. J. Gibson. But Bergson's model was embedded, and only worked within, a quite different model of time, a time that inherently demanded flow, in fact a flow that is indivisible, or in different terms, non-differentiable. I became concerned with how this model squared with relativity – our chief model of time. Throwing myself into the study of the subject, diving into the physics textbooks, I found myself deeply lost in the question of the status of the "effects" supposedly explained by relativity, the mysteries of the twin paradox, the reality (or not) of the frozen space-time "block" of all events – past-present-future – seemingly implied by the theory.

I wrote a paper, circa 1974, attempting to reconcile Bergson, perception and relativity. Shaw gave it to Gibson who happened to be visiting the University of Minnesota at the time. Gibson made a few comments on the paper, lauding me for taking on big problems, but basically he was having none of this "reconciliation." In a talk the next day at the university, he made the obscure statement that you will see quoted at the beginning of the introduction of this work, I'm sure coming out of the blue to everyone in the audience, i.e., everyone who had not written the particular paper on relativity that Gibson had just read. The statement was simple – and profound. I knew instantly what was meant. I had seen the same thing in Bergson, in his little work named *Duration and Simultaneity*, in which he attempted to cut through the confusion and set the interpretation of relativity and physics straight. This book is my effort to do the same.

Introduction: The Paradox

Physicists mislead us when they say
there is no simultaneity. When the camera pans
to the heroine tied to the rails and then to the hero rushing to the
rescue on his horse – these events are simultaneous.

— James J. Gibson[1]

A Paradox - Truly

In 1911, at the Congress of Bologna, the physicist, Paul Langevin, announced the "twin paradox" – to him, a remarkable implication of Einstein's Special Relativity.[2] As is known by virtually all interested in physics and in the subject of time, in the paradox, a twin boards a rocket and travelling at high speed – near the velocity of light – heads for star realms yet unknown. His brother is left on earth, playing solitaire. Well, Langevin didn't mention solitaire, but the earthbound brother has to be doing something. In any case, when his rocket riding brother returns, behold, the rocket-brother yet looks like Brad Pitt at twenty. The earthbound twin is sitting in a wheel chair, has grey hair, wrinkles, a long grey beard and arthritis in his hands from playing incessant solitaire. In a word, according to Langevin, the vast, near-light speed of the rocket had, according to the time changes described by the equations of relativity, caused the rocket-riding twin to age far more slowly than his stationary brother.

The remarkable implication here in Langevin's little story is that the changes of time described in relativity are really real, i.e., they have what is termed *ontological status*. The wrinkles, grey hair and arthritis of the earthbound brother are very real, while on the other hand, the vibrant youthful state of the rocket twin's physiology is also very real. The differential aging is a real effect. The time-changes are real effects. The time-changes have ontological status.

This judgment on the ontological status of relativistic changes of time has been embedded in the discourse on relativity ever since. Einstein agreed – well, yes and no, as we shall see. Physics used and uses the Lorentz equation with its ascription of a new "time" (t') to a moving object such as the rocket-riding twin to also explain the increased life-span (or slowing "aging") of a meson moving at high velocity as opposed to its more quickly dissolving life when at rest. The same equation too is used to explain the retarded or slowed down movement of a clock sitting on a jet where the jet is rapidly flying around the earth. Compared to a clock left at the airport, on return, just like the slower aging rocket-twin, the jet-carried clock shows that its time reading is a little late or retarded – its "hand" was moving more slowly.

But if we are talking about remarkable implications, then the truly remarkable implication of Langevin's paradox was that it actually completely destroyed the logical consistency of Special Relativity. Even more remarkable,

Einstein, in agreeing (again, yes and no) with Langevin, agreed with this destruction. In the wake of this destruction, the notion of the "space-time block" supposedly implied by relativity where all time is laid out – past, present, future – in a vast unchanging "block" was cemented and subsequently described by innumerable physicist writers, science writers, philosophers, and just plain writers. In this, yes, again remarkable phenomenon, we have the spectacle of numerous theorists working hard to explain, given the frozen, unchanging space-time block, how our "illusion" of the passage of time arises or our perceived motion of the flow of events. It has proven one of the most difficult of illusions to explain.

Not too long afterwards, along would come quantum mechanics, a theory that itself moved more closely towards the phenomenon of time as for example embodied in the uncertainty principle and its implications for the time-related phenomenon of motion, or in the Schrödinger equation describing the evolution of a system of particles. The two – relativity and quantum mechanics – prove to be virtually unmarriageable. Michio Kaku, in his public musings, argued that general relativity cannot be reconciled with quantum mechanics, stating flatly that one of these theories has to be wrong.

Einstein Meets Bergson

One of the most prominent thinkers of the era was deeply disturbed by what he saw and things he saw coming in the future given the acceptance of Langevin. This was the great French philosopher of time and mind, Henri Bergson, author of *Time and Free Will* (1889), *Matter and Memory* (1896) and

Creative Evolution (1907). His popularity had been great, the French newspapers even wondering if he should move his lectures to the Paris opera hall in order to accommodate the many ladies of society who wished to hear the lectures of the great Bergson.[3] The French forces in World War I had claimed his key concept in *Creative Evolution*, the "Élan Vital" or vital force – a concept far from the "vitalism" that is routinely disparaged by current philosophers and evolutionists – as the motto of the French army (which had the élan vital). The academic resentment towards him for his very popularity was

Figure I.1. Henri Bergson

so strong that he had decided to retire early from his academic position. But Langevin's implication was so disturbing, so problematic, it moved Bergson to emerge, if only briefly, from his retirement.

In 1922, he wrote a book, *Duration and Simultaneity*, on the subject, analyzing special relativity in detail, examining its implications, deriving and dissecting the equations himself – he was no mathematical slouch, having won a prestigious French prize in mathematics in his teens. In his words:

> To get at this [the true meaning of relativity], we went over the Lorentz formulae term by term, seeking the concrete reality, the perceived or perceptible thing, to which each term corresponded. This examination gave a quite unexpected result. Not only did Einstein's theses no longer appear to con-tradict the natural belief in a single, universal time, but they even corroborated it....[4]

The same year, 1922, Bergson engaged with Einstein in a spontaneous discussion under the auspices of the Société de Philosophie.[5] Acquiescing to an invitation to make an impromptu comment, Bergson noted, in the course of about fifteen minutes of remarks, that the concept of universal time arises from our own "proper" or experienced time in our immediate environment. He drew attention to the concept of the *simultaneity of flows*. Our experience of simultaneity, he observed, arises from our experience of multiple flows within a single flow, whether it be (using my own examples) multiple race cars racing side by side down the track, multiple melody lines within a single flow of a symphony, multiple musicians playing on the symphony stage, multiple women cooking in the kitchen, multiple family members eating at the table, a boat floating down a river with geese flying overhead, or Gibson's hero coming to the rescue of a struggling heroine. This experience of multiple simultaneous flows within a single experienced flow is generalized to other perceivers, ultimately, he argued, to our concept of a *universal* flow of time. Further, this intuitive notion of simultaneity supports the very concept of relating an event to a specific time instant on a clock (as for example where an observer must relate a lightning bolt and a clock hand at 3PM as occurring simultaneously). Now, he noted, a microbe observer could say to our observer that these two events (clock hand at 3PM, lightning bolt strike) are not "neighboring" events at all, but are vastly distant and would not be simultaneous to a moving microbe observer. Nevertheless, to paraphrase his conclusion, he felt that this intuitive simultaneity must underlie the possibility of any time measurement at all in relativity, and was in fact the basis for reconciling the two notions.

Einstein's reply is worthy of complete quote:

> The question is therefore posed as follows: is the time of the philosopher the same as that of the physicist? The time of the philosopher is both physical and psychological at once; now, physical time can be derived from consciousness. Originally individuals have the notion of simultaneity of perception; they can hence understand each other and agree about certain things they perceive; this is a first step towards objective reality. But there are objective events independent of individuals, and from the simultaneity of perceptions one passes to that of events themselves. In fact, that simultaneity led for a long time to no contradiction [is] due to the high propagational velocity of light. The concept of simultaneity therefore passed from perceptions to objects. To deduce a temporal order in events from this is but a short step, and instinct accomplished it. But nothing in our minds permits us to conclude to the simultaneity of events, for the latter are only mental constructions, logical beings. Hence there is no philosophers time; there is only a psychological time different from that of the physicist.[6]

This was the totality of the interchange. And so it rests. Bergson's position is, to say the least, a minority opinion. Einstein's "time of the physicist" has been the accepted criterion of reality. The simultaneity of perception is considered at best suspect and in practice, invalid.

Harold Stein, a philosopher of science, essentially reprised and expanded Einstein's argument, attempting to explain ongoing misconceptions of relativity, as he saw them, in terms of our continued naïve belief in the perception of simultaneous events – an illusion based on the high velocity of light. Thus, he argued in essence, the naïve or intuitive simultaneity that perception provides is founded upon the "fleeting motions" of "masses of elements" in the brain, all subject to the limitation of communication via the velocity of light, and implying therefore that at a small enough scale of time, perceptive simultaneity would break down.[7]

So we come again, in Stein, to consciousness and the flow of events in time. This is curious given that the problem of consciousness has now emerged as one of the great problems of the day – in physics, in philosophy, in cognitive science. It is a problem become ever more acute, far more so than realized in Einstein's time and even just becoming so in Stein's time. Neither Stein nor Einstein could claim to have a solution. Curiously, Bergson had such a solution. It is a solution that intrinsically relies on the simultaneity of events and the real flow of time. What if, we can wonder, the simultaneity of events is far from having been overthrown? What if the flow of time is far from having been locked in the block-time of relativity where all events are supposedly laid out – past-present-future – frozen? It is a solution, too, that generates a prediction that stands in contradiction to Special Relativity, but it is a contradiction if and only if physics holds that the relativization of simultaneity is a *real* property of time, i.e., a real, or ontological property of the matter-field and its temporal evolution. But this is the problem.

Introduction: End Notes and References

1. Gibson, the highly respected theorist of perception, made this statement in a talk at the University of Minnesota in 1975. He had read a paper by the author the previous day which at the time accepted Capek's (1966) view that relativity adequately preserves the "becoming" of the universe, and which attempted to fold in psychological time as part of the relativistic structure of time. Gibson, however, correctly appeared to have none of this, i.e., none of relativity whatsoever. He is in effect alluding to the concept of the *simultaneity of flows* of time, a subject discussed at length by Bergson in *Duration and Simultaneity* (1922/1965) in his analysis of relativity.
    Capek, M. (1966). Time in relativity theory: Arguments for a philosophy of becoming. In J. T. Fraser (Ed.), *The Voices of Time.* New York: Brasiller.
2. Later published as L'Evolution de l'espace et du temps, *Revue de Metaphysique et de Morale*, (1911), XIX, pp. 455-466.
3. Gunter, P. A. Y. (1969). *Bergson and the Evolution of Physics.* University of Tennessee Press.
4. Bergson, H. (1923). *Duration and Simultaniety*, New York: Bobs-Merrill, p. 5. (First published in 1922).
5. Gunter, P. A. Y. (1969). *Bergson and the Evolution of Physics.* University of Tennessee Press., pp. 123-135.
6. Gunter, P. A. Y. (1969). *Bergson and the Evolution of Physics.* University of Tennessee Press, p. 133.
7. Stein, H. (1991). On relativity theory and openness of the future. *Philosophy of Science*, *58*, 147-167.

CHAPTER I

The Strange Status of Relativistic Effects

It was then that I realized why I had been confused.
So long as I could imagine the time dilation
and other effects actually happening and
could work out the quantities involved,
that was all that was needed.

— Paul Davies, *The Matter Myth*[1]

Aging and duration belong to the order of quality.
No work of analysis can resolve them
into pure quantity.

— Bergson, *Duration and Simultaneity*

The "Koan" of Relativistic Effects

Let me begin with an overview of the problematic status of physical effects assigned to STR. It is a difficult topic, one which faces every student of the subject. It is in fact one of the worst of "koans," a "sound of one hand clapping" that boggles the mind, for the contradictions that arise in its interpretation are legion.

Relativity, it is well known, contains a feature which sees space units contracting and time units expanding depending on the motion of an observer. The most famous example, as we discussed in the introduction, is the twin paradox. In this case, twin Y leaves the earth at high speed in a rocket while his brother, twin X, stays on the earth. X is considered the stationary twin; he is at rest relative to Y. In motion at high velocity, Y's units of time, according to relativity, expand. Simultaneously, his space units contract. Because his time units are so much larger, he uses fewer of them, and when he returns to earth, he has aged less than his brother X. In this paradox, then, the expansion of time units and contraction of space units is considered very real. If the earth-based twin has a long beard, grey hair, and occupies a wheel chair, and the rocket-riding twin returns looking like Brad Pitt at twenty, well, we have a very real, a very physical, effect. These expansions and contractions, then, have *ontological status*, i.e., *actual being*. If this is the case, Einstein's "relativization of simultaneity" must be very real too.

What is the relativization of simultaneity? It relates to fundamental problems of measurement. Suppose, Einstein had argued, two lightning bolts strike on either side of you, fortunately a safe one thousand meters away. You happen to have two very accurate stop watches in either hand. Both are perfectly synchronized to the millisecond. You click to stop each of them when you see the light from each bolt out of the corner of your eye. You are a very fast and accurate "clicker." Behold, both watches show the same time. Further, you measure the distance from where you stood to the point

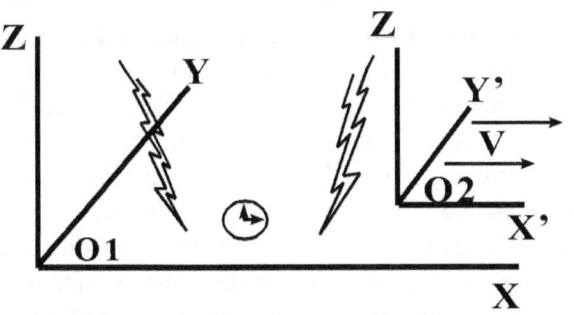

Figure 1.1. Two lightning bolts strike. Each strike at 3pm according to Observer 1 (O1) in the stationary system. Observer 2 (O2) is in the system moving at velocity, v.

2

where each bolt hit the ground. The distances are exactly equal. Assuming the light from each bolt traveled at the same velocity to your eyes, then the two bolts must have hit simultaneously. They traveled the same distance at the same speed, so they must have hit at the same time in order for you to have stopped both your watches at the same time. Therefore you judge these two lightening bolt events to be simultaneous. So far so good. But suppose another observer, we'll call him Observer Two, is moving on a large flying disc (his reference system) at some velocity right past where you stand (Figure 1.1). Observer Two is moving on an exact line towards the bolt on your right and away from the bolt on your left. He too has two synchronized stop watches. Note, however, that for this moving observer, the light from the bolt on the right must strike him a little sooner since he is traveling towards it, while the light from the bolt on the left gets to him a little later since he is moving away from it. He stops his two watches at different times. He declares the two-lightening bolt events *not* simultaneous.

Surely, we ask, he must know that he is moving! This explains the difference easily. But, said Einstein, perhaps he does not know that he is moving. Perhaps he thinks he is at rest. Perhaps he really is at rest. Perhaps it is you who are moving. How do we know? This became the essence of the first of two major postulates proposed by Einstein and which underpin his theory. The postulate is stated as, "the laws of physics are the same (invariant) in all inertial (reference) frames." It can equally be called the "reciprocity of reference systems." It implies that any observer has the right to declare himself at rest and all others in motion with respect to him. There is no way to tell who is right. The second postulate is the invariance of the velocity of light in all inertial frames.

Where do the expanded time units and contracted space units come from? Well, since Observer Two doesn't realize he is in motion (according to you), his clocks are not actually in sync. The method by which he must synchronize his clocks, Einstein showed, would be affected by his motion. One of his clocks will lag behind the other. Because of this, his measurements of distance and time within his own system will be affected. Einstein derived equations to allow us, as Observer One, to coordinate Observer Two's measurements of distances and times to our measures, in fact to specify what his measurements will look like in his system in terms of distance and time values. Central to the equations is a constant for both systems – the velocity of light. Applying these equations to Observer Two and his reference system, we would assign him expanded time units relative to ours. We would also assign him contracted distance units. At this point, one can intuitively understand why these distance and time change phenomena might be called "measurement differences." They are seeming squabbles over clock settings due to motion, but the problem of just who is in motion is very real. Observer Two, invoking reciprocity and declaring himself to be the system "at rest," can of course use the same equations for our system and for our distance and time values, claiming we are in motion and our clocks are out of sync.

Note what this implies for the simultaneity of events. The strikes of the two lightning bolts are relativized. They happen at the same time for one observer, at different times for another. Events that seem simultaneous to us may not be for another person. This means that what are simultaneous events for one observer may be successive events for another. This is to say, drilling down, that two simultaneous events for one observer, may, for another, be one event in his future, the other in his past. But what does this mean for the flow of time?

What is the classical conception of time? The advance of time traditionally involved the vision of the "time-growth" of the universe along some universally defined plane we call the "universal present." Were we to build a "space-time solid" in three-dimensions, letting the third dimension represent time, we could build one with (very thin) bread slices. Each slice represents all of 3-D space taken at an instant in time. We proceed, adding slice by slice to the "front end," gradually building a time-solid "loaf." The universal present is reduced rather mundanely to a slice of bread in this exercise. The flat surface of each slice is the universal "plane" of the present. In the classical conception, everyone's "present" is on this plane. All simultaneous events live on this plane. To us, the two lightning bolt strikes were on this plane. Any event not on this plane is either in the past, or the future – for all beings.

But now we have the relativistic fact that what are simultaneous events for one observer might be successive events for another. This implies different planes of simultaneity. It can be visualized as slices at different angles through our time-loaf. For observer X, with a plane sliced at a certain angle (Figure 1.2), certain events which he is experiencing as simultaneous events comprising his "present" can yet lie in the future for observer Y, while others lie in Y's past.

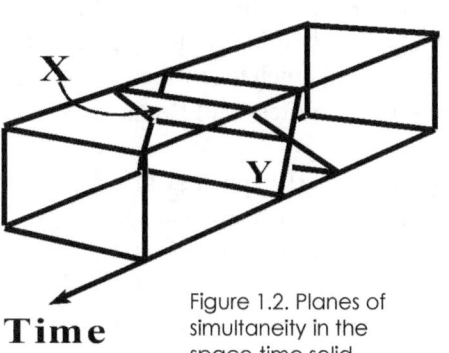

Figure 1.2. Planes of simultaneity in the space-time solid.

This vision of different futures and pasts for observers moving relative to one another makes it extremely difficult to conceive of a "universal becoming," with its vision of the growth of the universe in time along the plane of the "universal present." The conversion of simultaneities to successions, and successive events to simultaneous events, presents a troublesome difficulty for this classical conception, for the "plane of the universal present" seems to have disappeared – a single vertical slice cannot properly represent the "present."

There is, however, a natural route out of this dilemma, and it is simply to deny that there is any universal becoming, any motion of time, and to move instead to a conception of a static universe. Einstein's great collaborator, the mathematician Herman Minkowski, made statements that were the most famously conducive to this view. "Henceforth space by itself, and time by itself, are doomed to fade away into mere shadows, and only a kind of union of the two will preserve an independent reality." This conception is commonly called the "block universe." In it, there is no motion of time. All is given, past, present, future, in one giant block. This is a very common interpretation of relativistic space-time.

But let us remember, the *ontological* reality of this static block model entirely depends on the relativity of simultaneity being a fact. All depends on this relativization being a real property of the time-evolution (which we can no longer coherently visualize) of the matter-field. On this in turn depends the reality of the expanded time intervals and contracted space intervals of the rocket-riding twin Y. On this, in its turn, depends the differential aging of the twins X and Y, or the retarded aging of twin Y, as a real, physical property of matter, and the grey beards and real wrinkles.

Space Changes as Non-Ontological

When one begins to study the special theory, this is the first question that arises: are the changes in time and space real? It is extremely perplexing, for there is much to say that they are not real, and much to say they are. Here is a comment by the prolific physicist and physics writer, Paul Davies:

> How could the same thing [aging] happen at different rates?' I asked myself. I formed the impression that speed somehow distorts clock rates, so that the time dilation was some sort of illusion – an *apparent* rather than a *real* effect. I kept wanting to ask which twin experienced real time and

which was deluded. ... I had to admit I could not visualize time running at two different rates and I took this to mean that I did not understand the theory. ...It was then that I realized why I had been confused. So long as I could imagine the time dilation and other effects actually happening and *could work out the quantities involved, that was all that was needed.*[2]

It is not comforting to see the mechanical resolution he finally accepts, simply "doing the equations." But the contradictions are deep. Consider the initial and critical experiment to which the theory was applied, the famous 1895 experiment of Michelson and Morley. Michelson and Morley were trying to ascertain the speed of the earth through the ether. The ether was considered the all pervading, universal, fluid-like substance or medium through which energy is transmitted. Energy was considered to be propagated in waves. A wave requires some medium to ripple, in fact a wave is simply a ripple propagating through the medium. Without something like the ether, there could be no waves of energy. The earth was conceived as though it were a huge boat plowing through the ether, creating a bow wave or current. The Michelson-Morley experimental apparatus

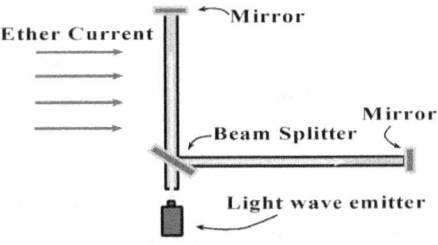

Figure 1.3. The Michelson-Morley apparatus (1895). The earth was conceived as a boat plowing thru the ether, creating an ether current or flow. The pipes/arms of the apparatus are equal in length, and an emitted light wave is split in both directions. The light wave traveling through the pipe in the direction of the current and back should have taken a little longer time, creating an interference pattern or fringe between the two waves (rather than the two waves meeting precisely at the same time, creating no interference). However, no interference was observed; each wave takes the same time, creating a problem for the existence of the ether.

(Figure 1.3) sent out two light waves at right angles to each other. One went against the current, one went crosswise to the current.

When they ran their experiment, they obtained a strange result. The light ray running in the direction of the ether current and back should have taken longer than the light ray running crosswise. It did not; both rays took equal times. The result could be explained if the arm of the apparatus, in the direction of motion, in line with the ether flow, shrunk slightly, just enough to compensate for the theoretically larger time of travel of the light ray going though it. The light ray cheats by having a shorter course. Is such a contraction of the arm of the Michelson-Morley apparatus real, a physical fact?

Let us remember that Hendrik Lorentz, a highly respected physicist of the time, some years before Einstein's publication, originally proposed that it was indeed real. He advanced ether-based, electro-dynamical arguments in support of equations he developed for the fore-shortening of the apparatus-arm in the direction of motion as a function of velocity. His equations expressed the degree of contraction and accounted for the same travel-times. The equations looked exactly like Einstein's. But the contraction was unappealing to physics; it was rejected, or at least never accepted. Why was Einstein's "contraction," using precisely the same equations, accepted? Because the length became a space-time invariant.

How does the length become such an invariant? By being subject to the reciprocal transformations of two observers in two different reference systems, either of which can consider himself at rest and the other in motion, and this in conjunction with the joint compensation of increased time intervals for decreased space intervals (or units of measure). Einstein's perceived advance was to embed the Lorentz transformations within this symmetric, reciprocal framework, together with postulating the invariance of the velocity of light. There was no longer a "length in and of itself, rather a space-time "length" composed of varying amounts of space and time, the total, shall we say, being an invariant. Indeed, Einstein wished that his theory had been named "Invariantentheorie," rather than relativity.[3] In special relativity, the Lorentz transformations have no meaning with respect to just one observer. There is no invariance with just one observer. *Some form of transformation is required for an invariant.* This symmetric system is required, and within it, either observer can declare himself at rest, and then attribute the length contraction to the other (who is in motion), adjusting the other's space and time units to preserve the invariance of the velocity of light. Therefore as the physicist and M.I.T. professor, A. P. French, states in his textbook on relativity, the length contraction is not a real property of matter, *it is a measurement effect*, "something inherent in the measurement process."[4]

In the textbooks I studied in the 1970s, the explanations of length contraction routinely told this story. The length contraction is not real. It is an effect of measurement only. The length is a space-time invariant, but no single observer has a claim on knowing the "true length." The student is warned not to fall into "the length contraction is real" trap. In truth, we must remember, there is little choice. To say that it is a real effect is to say that the Michelson-Morley apparatus arm is actually contracting somehow. This is to revert back to Lorentz and his hypothesized contraction, an explanation in fact with a real, physical model at its base – the very thing physics refused previously to accept.

Time Changes as Ontological

But as soon as the textbook turned to expanded time units or time dilation, the story was different. The problem was that there were real, physical phenomena for which time dilation appeared to be physics' only available explanation. Mesons, for example, are particles that have a certain lifespan. At rest, they exist for a certain measurable period before they decay away. When moving at high velocity, they exist for a longer period. When Lorentz's original equations are applied in this case, the increased time is perfectly predicted. Therefore time dilation is considered a quite real effect.

If there is a doubt that this is considered a very real effect, we can propose a test. We could set up a tiny electric switch a distance from the start of the meson's motion. The distance is just long enough that if the electron is not living any longer beyond it normal rest life, it won't set off the switch, but if it is living longer, it makes it to the switch and sets off an alarm clock. The ringing clock is a very real effect. Physics would quite surely accept that the meson will ring the clock.

The slow-aging Y twin with the grey and bearded X twin is simply another case of the time-dilation being considered a real effect. There are two problems with all this. The first is that it ignores the reciprocity of reference systems. A tiny physicist on the meson should be able to say, "I'm not in motion, you are. I will never make the clock ring." The rocket-riding Y twin has perfect right to declare himself at rest, and the X twin in motion. The fact that he is on the rocket is of no account. The rocket engines could be considered to be holding the rocket's place in space as the earth moves away from the rocket, but in truth, the mathematics of relativity is abstract and these physical considerations are irrelevant. Only the abstract reciprocity of reference systems is important. So now it is the X twin who ages less. So for whom is the aging less? X or Y? Has *time* really changed? Or should we just be saying that aging period too is a space-time invariant, just as the length contraction? But fast forward. An experiment was ultimately performed in which a clock was put on a jet and flown at great speed. When the jet landed, the clock was compared to a previously synchronized counterpart left on the ground. The jet-carried clock lagged behind. The Lorentz equation for the expanded time-interval accounted for the difference – another triumph for relativity. When the experimenters stepped off the jet with their retarded clock, no one stepped forward and argued that in actuality the plane was at rest and the earth moving at extreme speed relative to the jet, thus it is the earth-based observers' clocks that should be retarded. Why not? Because obviously it is absurd. These are very real effects. They cannot be made to go away by invoking reciprocity. If the longer-living meson rings the alarm clock, the ringing is very real, it cannot be said that clock isn't ringing by suddenly remembering reciprocity. The bearded twin, should it happen, would be very real, and the beard would not go away by remembering reciprocity. The symmetry implied by reciprocity clearly has been broken.

The second problem is simply that in the system in which Einstein embedded the Lorentz transformations, space changes compensate for time changes and vice versa, depending on the motions of the observers. This is why there are only space-time invariants in Einstein's system. If space changes are simply compensating for time changes, there is no way the two forms of changes can have differing ontological status – either both are ontological or both are not ontological, i.e., in the non-ontological case, they are simply measurement effects. As relativity was accepted precisely on the basis that Einstein showed the Michelson-Morley result to be a measurement effect, that the apparatus are was not actually, physically foreshortening in the direction of the ether motion as Lorentz had suggested, then physics accepted space-changes as non-ontological, inescapably then including time changes as non-ontological. This should end the story on the reality of the "changes in time," period. Unbelievably, pathetically, it did not. We continue.

Space Changes as Non-Ontological – Again

As far as I can ascertain, in the 1980s (perhaps earlier) another paradox began appearing in the textbooks called the "pole-barn" paradox (Figure 1.4). The "paradox" notion was now being applied to the length contraction. In this paradox, we have a longish, say, telephone pole. In its resting state, it is too long to fit into a certain barn. However, when the pole is launched into motion at a velocity near the speed of light and flies through the barn, there is a period where the pole, due its length contraction, actually fits into the barn. But *this* paradox is used as a parable for illustrating that we should *not* consider these real effects. It is unhesitatingly pointed out that the *barn* could be conceived to be in motion, and therefore the barn will contract. Now the pole does *not* fit. So the length contractions are not real, or in philosophical terms, they have no ontological status. This nicely holds the line with the interpretation of the Michelson-Morley experiment.

One could ask something however. Just like the jet-carried clock experiment, why not perform a pole-barn experiment? We could rig a mini barn-like apparatus with front-end and back-end doors that open and shut at great speed, or some analogy. The device would capture a mini-pole moving at high velocity precisely when it fits inside due to its length contraction. If we can so unhesitatingly predict that the jet-carried clock will slow down, why would we not predict that the mini-pole would contract and be trapped in the barn? But this would be admitting that the length contraction too is a very real effect. It would signal the end of any pretense of usage of the reciprocity of reference systems aspect of the special theory. At present, physics deploys the reciprocity feature for length contractions, and unhesitatingly dumps the feature for time-expansion. It therefore rejects the relativization of simultaneity as real and simultaneously (or not simultaneously?) accepts the relativization of simultaneity as real along with its block universe implication.

Those knowledgeable in this area may say, "But the twin paradox must be assigned to the General Theory (GTR)." This is due, it is thought (by some), to the accelerations involved with the rocket. Einstein's General Theory, developed after STR, deals with gravity and acceleration. This is an obviously questionable assertion on face value. If it is the twin's beard, i.e., the real, physical, obviously non-symmetric effect displayed in the aging that we are worried about, then the jet-carried clock and the meson's increased life spans must be sent to the GTR as well. These are just as real and just as non-symmetric. In other words, all such effects must be moved into GTR, leaving the explanatory effectiveness of STR a blank. But I will

Figure 1.4. The Pole and the Barn. At high velocity, the pole fits inside the barn. At rest, it does not fit.

deal with this later. Suffice it to say for now that this gambit only adds to the confusion. One quickly discovers that there is an "explanatory pea" being shuffled between the General Theory and the Special Theory.

Chapter I: End Notes and References

1. Davies, P. & Gribbons, J. (1992). *The Matter Myth.* New York: Simon & Schuster, p. 101.
2. Davies, P. & Gribbons, J. (1992). *The Matter Myth.* New York: Simon & Schuster, p. 100-101.
3. Horton, G. (2000). *Einstein, History and Other Passions.* Cambridge, Massachusetts: Harvard University Press.
4. French, A. P. (1968). *Special Relativity.* New York: Norton, p. 114.

CHAPTER II

Special Relativity and Perception

We may not be able to say
what parts of the whole are in motion, motion there is in the
whole nonetheless.

— Bergson, *Matter and Memory*[1]

The Question for the Problem of Consciousness

Already a theory of consciousness has appeared by John Smythies, as we shall see, that assumes the standard vision of the implications of special relativity for time, namely that of the space-time block.[2] Weyl, a physicist contemporary of Einstein, expresses the implications of space-time unambiguously:

> The scene of action of reality is not a three-dimensional Euclidean space, but rather a *four-dimensional world, in which space and time are linked together indissolubly.* However deep the chasm may be that separates the intuitive nature of space from that of time in our experience, nothing of this qualitative difference enters into the objective world which physics attempts to crystallize out of direct experience. *… Only the consciousness that passes on in one portion of this world experiences the detached piece which comes to meet it and passes behind it,* as *history…*[3]

Weyl's statement, implying that the experienced passage of time has no objective counterpart, would have had revolutionary implications had it truly been taken to heart. But relativists themselves do not seem to have been entirely clear on the implications of the concept of space-time, and the meaning of these statements had perhaps more radical ramifications than anyone cared to make clear to anyone. We will briefly examine these below.

We are about to spend some time here on consciousness and perception as they relate to relativity. Both are intrinsically related to the subject of the flow of time, the precise subject at issue here in questioning the reality of the "space-time block" and, as well, the reality of the "changes of time" supposedly implied by relativity. Further, this topic begins to introduce us to the profound basis for the "simultaneity of flows" referred to by Gibson, and the ultimate inability of STR to address this.

The 'Psychical' Observer

The extensions of time-extended objects are usually called "world-lines" in relativity theory, or sometimes "tracks." "An individual," says Eddington, "is a four-dimensional object of greatly elongated form. In ordinary language, we say that he has considerable extension in time and insignificant extension in space. Practically, he is represented by a line – his track through the world". [4] The last five words – "his track through the world" – as Dunne (1927) pointed out, make his statement appear like hedging, for we must ask how the line can be both the observer and the observer's path.[5] But Eddington makes clear within the same page that the track is indeed coincident with the observer, i.e., is the observer himself. "A natural body," he says, "extends in time as well as space, and is therefore four-dimensional."

Now the first problem that presents itself is the experience of the passage of time that humanity universally shares. If everything is given, if the universe simply exists as a four-dimensional, static block of space-time, then motion has become non-existent. "Changes then correspond to individuals moving along world-lines" – this is the acknowledgment of our experience of time's motion. But just what are these individuals? To any observer viewing such a system of fixed tracks or world-lines, the appearance of motion in the dimensions representing space could be produced by the movement of a three-dimensional field of observation along a track or fourth dimension orthogonal to the other three. Thus the field would simply "come across" events (as does the 1-D field of Figure 2.1). This time-traveling field of observation we can provisionally term a "psychical" observer, for the physical observer is defined as the track traveled over. This is exactly the move Smythies accepted and utilized, envisioning "consciousness modules" moving along these tracks.

The relativists had a complex case to present, and the burden of a psychical observer, had it explicitly been acknowledged, would probably have been too much to bear. Not wanting to ignore the motion of time, however, at least some expositors of this particular notion of space-time leave us with

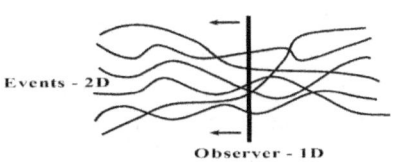

Figure 2.1. One-dimensional field traversing events in a 2-D universe

the non-committal statement indicating that the observer moves along his track, from which the reader may infer what he pleases. The reader usually proceeds to infer that the observer is nothing more than an organic, physical apparatus, and that this physical apparatus moves over its nebulous track in the fourth dimension. Obviously, however, a track that possessed reality to such an extent as to account for the physical characteristics of an imagined 3-D object moving along it would be, in every one of its cross-sections, physically indistinguishable from the object. Physically the track is the object extended four-dimensionally. Anything which we would consider moving along the track must differ from the track itself. Speaking of a body such as a clock or light ray moving over its track is conducive only to confusion, for the clock is physically a bundle of tracks and cannot move over itself.

Some philosophers, such as J.J.C. Smart, have noted this inconsistency.[6] Yet, respecting the static, "all is given" nature of the four-dimensional manifold, have voted solidly in favor of the concept that "there is no time." They see the passage of time as a pure illusion. Unfortunately, while they scoff at the absurdity of a psychical observer or of "consciousness running along world-lines," they offer little to put in its place. *You must at least offer a "theory of the illusion."* Even while Smart is writing his essays on time, his hand fatiguing, the ideas flowing by, he is experiencing the "illusion" in all its trickery. Whence then does the experience of the "passage" of time arise? At least the admittedly mysterious psychical observer tried to answer the question.

The "Specialness" of Moments

Nevertheless, the boldness with which relativistic writers have announced the reality of the "block" and reconciled themselves to the ubiquitous "illusion" of the passage of time has steadily increased. A recent one by Andrew Thomas (*Hidden in Plain Sight,* 2013), unreservedly endorses the "block" concept (as do nearly all expositions), then, venturing into the problem of consciousness and the now needed explanation of the illusion of the passage of time, simply claims that each moment in experience is "special" – my day in 1952 was special, my day in 1985 was special, my day in 2013 was special – so the specialness of each experience makes us feel time is passing, when in fact all such experiences are "special" simultaneously in block time. This too is absurd; we shall see this ever more deeply as we discuss the nature of the classic metaphysic and its abstract time below, but for now we can simply consider this: I experience a leaf falling, twisting to the ground; I experience a fly buzzing by. These are experienced as flows. What then is the duration of each special moment in such flows? Since we are trying to explain the experience of flow by the "specialness" of each moment or instant along a static world-line in the block, then each "instant" in such a flow must be "special," and with not much effort, we would arrive at the logical consequence that the duration of each "special" instant could only be that of a mathematical point, i.e., no duration, therefore, no flow. Thus we are back to a timeless, unchanging block where each mathematical point/instant along a world-line is "special" – there is no explanation whatsoever of the experience of flow.

One will see a (confused) response to these points to the effect that, "This is just like watching a series of static film frames, yet experiencing flow." Ignoring the problem in this response wherein it neglects the fact that the static, timeless points to which this (non)flow (of actual events in time) has to be reduced now can look nothing like the world at our scale, namely "buzzing" flies and "twisting-falling" leaves, what is surreptiously assumed here is this – our consciousness that itself is the flow and to boot, the fundamental memory that provides continuity. The static film frames notion is cousin to the standard notion that perception (the brain) is taking snapshots of an ongoing event, say a fly buzzing by, and using these static frames or snapshots to register motion. Yet how are we accounting for the experience of motion? Why is each snapshot not simply experienced, then lost as the next arrives, i.e., nothing is ever in experience but a static picture – a frozen fly in one position – then another, then another? What is providing the continuity such that a motion is experienced? Are we storing each snapshot in the brain? Then we have the equivalent of a set of photos of the buzzing fly laid out on a desktop – a set of static pictures, i.e., a set of immobilities. Do we now invoke some internal "scanner" sweeping across the photos to account for the motion? Then how does the scanner perceive motion? We begin an infinite regress. All this touches on a profound set of problems of perception, the relation of mind to time, and the nature of time itself, ultimately resting at the heart of Chalmers' "hard problem," a difficulty we will soon be addressing.

A Scale-less Manifold

There is yet another thing, already touched on above, for we have no right to assign any particular time-scale to this manifold. We cannot envision it as it would appear to normal perception, for this perception already entails a summation over a vast history of events. If the event/world-lines the psychical observer is passing over comprise a "buzzing" fly, the choice of scales is infinite. The fly can be merely a phase in a field of vibrating strings, an ensemble of electrons/protons with no precise boundary, a fly slowly flapping his wings, or the buzzing fly of our normal perception. We would then have to account for the means whereby the time-traveling field determines scales.

Smythies would envision his traveling consciousness module as projecting a camera-like mechanism into the brain, observing the brain-tracks.[7] Again, what scale is the "camera" observing – quarks, molecular activity, chemical flows? And how are any of these – quarks or whatever – unfolded into the world of golf balls and putting greens? This is simply what I have termed elsewhere the "coding problem."[8] A code, say three dots (...) can stand for, or be mapped to, an "S" in Morse code, the three blind mice, or Da

Vinci's nose. The problem is knowing the domain that the code is mapped to. To unfold the neural code into which the external world has been "encoded," the "camera" would have to already know what the world looks like. How is the external world of golf balls and greens unfolded from this chemical/neural/atomic code? The contents of the tracks are supposedly projected on the consciousness module's "screen." Welcome to the homunculus (or tiny little man) observing the screen. Nor are we clear why we seem to have a whole set of observation fields moving along in parallel and constituting humanity. Why are some of us not now fighting the Peloponnesian Wars – or are we?

In any case, we could exhaust ourselves on the metaphysical, epistemological, and psychological facets of the static block reading of the implications of STR. Had psychology considered it seriously, an immediate question might have been: why are we storing memory in the brain? Clearly all events are preserved in the 4-D manifold, and the brain itself is vastly four-dimensional. If our psychic observer can go forwards, why not backwards too? Or is storage merely an illusion in the first place as we are merely coming across things that resemble past sections of the track, sections corresponding to remembering events? These and other questions might have occurred.

One might wonder how STR can pose any dilemma for a theory of consciousness when relativistic effects such as time dilation only occur at any appreciable magnitude at extremely high velocities. The normal motion velocities of organisms seem such as to make STR's effects irrelevant. However, the strange implications being noted here – the inability to account for the experienced motion of consciousness, the specter of "psychical" observers as a questionable solution to this, the curious questions about memory – are all simply functions of taking a static, four-dimensional block model of space-time seriously. This model in turn only has a possible reality if we take the relativity of simultaneity seriously (as did Smythies), i.e., as having ontological status. Proposed STR-effects such as the twin-effect, even though occurring at extremely high velocities, cement in the ontological status of these effects, and therefore the reality of the relativity of simultaneity. In the theory, we shall see, the breakup of simultaneity begins at the most minute of velocities. Further, as we will also see when reviewing the analysis of Hagan and Hirafuji, whether or not the changes are taken as ontological, if STR is indeed valid, it places difficult constraints upon any theory of consciousness. Finally, in any case and regardless of discrepant orders of velocity, the Bergson model of perception, which I will briefly describe, generates a testable prediction relative to action that contradicts an implication of STR, again, only if STR's effects are taken as ontological.

Let me state this emphatically: I am not denying the reality of increased life-spans of mesons, or retarded jet-carried clocks. These phenomena are very real. The crucial question is: *how they are explained?* If changes of space and time, *as currently explained by the mathematics of relativity,* are ontological, then the relativization of simultaneity must be real. We are forced to the static block universe. A theory of consciousness is then held by this constraint, despite the difficulties into which it would inevitably place psychological theory. Given all these immensely problematic and incomprehensible implications of the static block universe for a theory of consciousness, it is time to move to a different framework of thought on the subject. We shall now briefly view Bergson's solution to the problem of conscious perception, a solution that goes to the source of STR's problem.

Bergson and Time

Let us begin with the heart of the difference between Bergson and Einstein. The "microbes" in Bergson's comments are an index, in essence an index to the process of thought leading to the "objective" that Einstein must take to its logical conclusion. Bergson, in introducing them, had asked just what is the concept of "proximity" or "neighboring events" used in relativity to relate clocks to events? A microbe consciousness questions whether the clock and lightning bolt of the system of some observer are "neighboring." A micro-microbe questions the microbe's judgment of what is "neighboring"; a micro-micro-microbe does the same to the micro-microbe, and so on. Logically, we are forced to take this to its conclusion. There can be no accepted judgment of neighboring (and therefore of simultaneity) as we descend scales until we end at the mathematical point. The mathematical point is the essence of complete abstraction. The question is, is time found at all at this abstract point-event?

At the foundation of Bergson's theory was already a critique of the *abstract* space and time implied in Einstein's theory-to-be. Abstract space, Bergson argued, is derived from the world of separate "objects" gradually identified by our perception. It is an elementary process, for perception must partition the continuous field that surrounds the body into objects upon which the body can act – to throw a "rock," to hoist a "bottle of beer." This fundamental perceptual partition into "objects" and "motions" is reified and extended in thought. The separate "objects" in the field are refined to the notion of the continuum of points or positions. As an object moves across this continuum, as for example, my hand moving across the desk from point A to point B, it is conceived to describe a trajectory – a line – consisting of the points or positions it traverses. Each point momentarily occupied is conceived to correspond to an "instant" of time. Thus arises the notion of abstract time – the series of instants – itself simply another dimension of the abstract space. This space, argued Bergson, is in essence a "principle of infinite divisibility." Having convinced ourselves that this motion is adequately described by the line/trajectory the object traversed, we can break up the line (space) into as many points as we please. But the concept of motion this implies is inherently an infinite regress. To account for the motion, we must, between each pair of points supposedly successively occupied by the object, re-introduce the motion, hence a new (smaller) trajectory of static points – ad infinitum. It is the core of Zeno and his paradoxes.

Zeno, Bergson held, was forcing recognition of the logical implications of this infinitely divisible, abstract space and time. With each step, Achilles halves the distance between himself and the hare, but he never catches the hare; there is always a distance, no matter how minute, between pursuer and pursued. In the paradox of the arrow, the flying arrow occupies, at each instant, a static point in space, therefore, "it never moves." In all four of the paradoxes, it is the infinitely divisible space traversed that is the focus. Motion, Bergson argued, must be treated as *indivisible*. We view the indivisible steps of Achilles through the lens of the abstract space traversed and then propose that each such distance can be successively halved – infinitely divided. Achilles, never reaches the hare. But Achilles moves in an indivisible motion; he indeed catches the hare.[9]

But the abstraction is further rarified. The motions are now treated as relative, for we can move the object across the continuum or the continuum (or the coordinate system) beneath the object. Motion now becomes immobility dependent purely on perspective. All real, concrete motion of the universal field is now lost. *But there must be real motion.* Trees grow. People age. Stars grow cold. Galaxies collapse. Bergson would insist:

> Though we are free to attribute rest or motion to any
> material point taken by itself, it is nonetheless true that the
> aspect of the material universe changes, that the internal con-
> figuration of every real system varies, and that here we have
> no longer the choice between mobility and rest. Movement,
> whatever its inner nature, becomes an indisputable reality. We
> may not be able to say what parts of the whole are in motion,
> motion there is in the whole nonetheless.[10]

He would go on to note:

> Of what object, externally perceived, can it be said that it moves, of what other that it remains motionless? To put such a question is to admit the discontinuity established by common sense between objects independent of each other, having each its individuality, comparable to kinds of persons, is a valid distinction. For on the contrary hypothesis, the question would no longer be how are produced in given parts of matter changes of position, but how is effected in the whole a change of aspect."[11]

Within the global motion of this whole, the "motions" of "objects" now become *changes or transferences of state.* The motion of this whole, this "kaleidoscope" as Bergson called it, cannot be treated as a series of discrete states. Rather, Bergson would argue, this motion is better treated in terms of a melody, the "notes" of which permeate and interpenetrate each other, the current "note" being a reflection of the previous notes of the series, all forming an organic continuity, a "succession without distinction," a motion which is indivisible. In such a global motion, there is clearly simultaneity.

The process of "objectification" which Einstein, in his response to Bergson, describes and accepts as leading us to the "real," to objective events, and which leads Stein to his "fleeting motions" of masses of "elements," is exactly the process warned of by Bergson. The "objects" of perception – purely practical partitions carved by the body's perception in the flowing universal field at a particular scale of time – are reified into the concept of abstract, independent "objects" and their "motions," and this is further rarified to "objective" space and time, with its objective, separable "events." And following this path, Einstein is consistent. These "objective," separate events are only mental constructs. They and their simultaneity are fully subject to the relativity logically inherent in their birth.

Physics on the Abstraction

Hence, to Bergson, Einstein's "time of the physicist" is an artificial time. It can be argued, however, that this (artificial) path is exactly the opposite of what physics has found itself to be following. The concept of abstract space and time – this "projection frame" for thought originating in perception's need for practical action – has been the obscuring layer that is slowly being peeled away. As Bergson argued, "...a theory of matter is an attempt to find the reality hidden beneath ... customary images which are entirely relative to our needs ..." [12] The customary images are dissolving. The *trajectory* of a particle no longer exists in quantum mechanics. If attempting to determine through a series of measurements a series of instantaneous positions, simultaneously we renounce all grasp of the object's state of motion. In essence, as de Broglie (1947/1969) would note, the measurement is attempting to project the motion to a point in our abstract continuum, but in doing so, we have lost the motion. Motion cannot be treated as a series of "points," i.e., *immobilities*. Thus Bergson noted, over forty years before Heisenberg, "In space, there are only parts of space and at whatever point one considers the moving object, one will obtain only a position."[13]

Lynds (*Foundations of Physics Letters*, 2003), echoing Bergson, now argues that there is no precise, static instant in time underlying a dynamical physical process.[14] If there were such, motion and variation in all physical magnitudes would not be possible, as they (and the universe itself) would be frozen static at that precise instant and remain that way. Consequently, at no time is the position of a body (or edge, vertex, feature, etc.) or a physical magnitude precisely determined in an interval, no matter how small, as at no time is it not constantly changing and undetermined. The inherent *uncertainty* introduced by this unceasing flow of time is the inescapable tradeoff required for the universe to change. It is only the human observer, Lynds notes, an observer employing a static conceptual background – who imposes a precise instant in time upon a physical process.

To take this for a moment back to Zeno, as I noted earlier, when the instant is ultimately reduced (if ever possible) in this infinite division process to a mathematical point, the theory of motion in the metaphysic is shown to be incoherent. The mathematical point is the instant reduced to an indivisible extent. This instant does not have a start point and an end point between which the arrow can move. If it did, it could be divided – a start half and an end half – and it would not be indivisible. If time then is made up of indivisible instants, instants by definition in which there can be no motion of the arrow, how then does the arrow move? It can only do so by being at successive intermediate points at successive intermediate times, i.e., it never changes its position over an instant, but only over intervals composed of instants, by occupying different positions at different times. In Bergson's critique, "movement is composed of immobilities," i.e., an absurdity. But equally, this is the very condition of which Lynds speaks, for at such an instant, the universe would be frozen in time – all change is impossible. *It would require that the entirety of space be reborn*, or be re-generated, instant after instant. But this is simply yet another regress, the motion of the hidden process behind the regeneration of all of space now becoming the issue. The fact is, reprising Lynds argument, there can be no instant of time underlying a body's motion. If there were, it could not be in motion. Rather, at no time is the body or the evolution of the matter field itself not constantly changing, no matter how small the time interval. As such, at no time does the body have a *determined position*. There is not an instant of time underlying the motion of Zeno's arrow at which it would occupy just one block of space or one tiny length of its spatial trajectory – its position is constantly changing.

Thus Nottale, noting Feynman and Hibb's proof that the typical paths of quantum particles are continuous but non-differentiable, now questions the fundamental assumption that space-time is differentiable, laying out a fractal approach to space-time, i.e., indivisible extents.[15]Nottale is referencing the geodesics that describe the motions of all particles in space-time – obviously a huge, sweeping scope – and his theory treats these curves as fractal, i.e., they are irregular and constantly changing at every scale; there are no straight lines. In the awesome implications of the fractal world, "at every scale" means that anywhere one looks at one of these curves, at the most infinitesimal of scales, one will find an inflection point, hence these curves are non-differentiable – there is no velocity derivable at an instant. In the general non-mincing of words of Nottale, "space-time is non-differentiable." This is also an overturning – along with the critique of Bergson – of what has hitherto been a mainstay of the theory of motion in the classic metaphysic. The essence of differentiation is division, say of the slope of a triangle or of a motion from A to B, into successively smaller units, taking the "limit" of what is in fact an infinite operation, that is, employing a mathematical technique to arbitrarily cease the divisions. To state that space-time is non-differentiable another way, we may say the global evolution of the matter-field over time is non-differentiable; it cannot be treated as an infinitely divisible series of states.

A matter-field in a global motion, wherein the motions of objects are changes or transferences of state, implies a simultaneity of causal flows. It also implies a framework for the problem of perception.

The Classic Metaphysic and the Hard Problem

Abstract space and abstract time form what can be termed the "classic metaphysic." STR dwells solidly within this metaphysic; it is only a refinement of the metaphysic's implications. It is this metaphysic that resides behind the entire discussion of qualia and the "hard problem" of consciousness as so coined by David Chalmers.[16] One must explain, argued Chalmers, given some model of the brain as a neural architecture, or as a connectionist network, or as a computer architecture, how this architecture accounts for the "qualia" of the perceived world – the "redness" of the sunset, the taste of cauliflower, the greenness of the grass. The question has engendered massive debate in the philosophical literature, with numerous attempts at solutions, none of which work – for all discussion yet proceeds within the classic metaphysic and its abstract space and abstract time.

As noted, the end result of the "principle of infinite division" that this "space" represents, even could we legitimately conceive of an end of such an operation, ignoring the mathematical hand waving of taking a "limit," would be at best a mathematical point. At such a point, there could exist no motion, no evolution in time of the field. Further, as every spatially extended "object" is subject to this infinite decomposition throughout the continuum, then we end with a completely *homogeneous* field of mathematical points. The continuum of mathematical points then, both spatially and temporally, can have no qualities – qualities at the least imply heterogeneity.

That this is indeed the framework that the debate participants have tended to work within is attested to by a very common starting point, namely that the matter-field contains no qualities – objects have no color, there are no sounds, etc. This framework is also betrayed by the fact that the vast preponderance of examples of qualia are static – the "redness" of red, the taste of cauliflower, the feel of velvet, the smell of fresh cut grass. Seldom are qualities of *motions* ever discussed, e.g., the "twisting" of leaves, the "gyrations" of a wobbling, rotating cube, the "buzzing" of a fly. This glaring lack is coordinate with the fact that an abstract "time" that is simply another dimension of the infinitely divisible space is equally completely homogeneous. Any "motion" in this space, logically, has no duration greater than a mathematical point, then another point, then another... In fact, then, the debaters universally fail to realize that the perceived time-extents of these motions – the rotating cube, the buzzing fly, the whirling of the coffee surface with circling spoon – are equally *qualities* that arise, just as problematically as the "static" colors of objects, in the homogenous time dimension of infinitely divisible instants in this continuum.[17]

Galileo, in initiating this metaphysic, equated the *real* with the quantitative.[18] Qualities, he felt, were contributions of the "living organism." From this arose the distinction of primary and secondary properties of matter. Shape (form) is considered part of the quantitative realm and thus considered part of the "real," not a quality therefore and not part of the hard problem. But the concept of a static instant is a fiction. This is why Galileo was wrong when he assigned shape or form to his "quantitative" continuum, while thinking he was excluding qualities (contributions of the mind) there from.

Figure 2.2. Texture density gradient (say, the surface of a table top) with a cup in two different positions.

The current conception of the derivation of form is based on velocity flows. It is J. J. Gibson who had earlier pointed to the significance of texture gradients and these flows or optic flow fields. Figure 2.2 is an example of a texture gradient. The little circles are the "texture elements." The size of these elements and their horizontal separation (S) changes in perfect mathematical proportion with the distance from the eye, or $S \mu 1/D$. The vertical separation (S) between each element decreases in proportion to the square of the distance, $S \mu 1/D^2$. These texture gradients are ubiquitous – floors, beaches, lake surfaces, gravel driveways, tiles, etc. Turn the gradients upside down – you see them as ceilings or the bottom of clouds. For

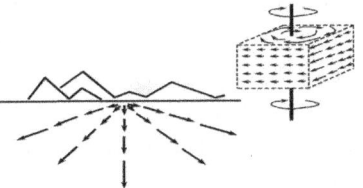

Figure 2.3. Optical flow field. A gradient of velocity vectors is created as an observer moves towards the mountains, with the speed of the vector (represented by the length) inversely proportional to the squarre of the distance form the observer ($v \mu 1/d^2$). The flow field "expands" as the observer moves. At right, the flow fields over the side of a rotating cube – expanding as the side rotates towards the observer, contracting as it rotates away, with the top a radial flow field.

Gibson, the information in these gradients specifies (or is "specific to") receding distance. If we think of the cups in the figure as the same cup in two different positions, or the cup in the background as moving to the foreground, the perceived constant size of the cup is specified by another constant ratio of proportion (or invariant), for the height of the cup increases as it moves such that it is inversely proportional to the number of rows of texture units it occludes, e.g., from four rows occluded in the background to two in the foreground.

We can set these gradients in motion, as when driving our car down the road. We now have an optical flow field (Figure 2.3).[19] It was the attempt to solve the "correspondence problem" that drove theorists on the perception of form to these flowing fields. To take a simple example, the correspondence problem visualized the rotation of a cube (or any form) as divided into a series of frames (or snapshots). To compute the cubical form over this motion, the features of the cube – its edges and vertices – have to be tracked from frame to frame, i.e., the *identity* of each feature must be maintained/tracked from frame to frame. But this correspondence problem – tracking all the features – was eventually considered intractable. Thus current perception theory sees perceived form

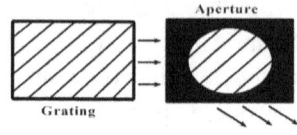

Figure 2.4. The aperture problem. The card with the grating is moving to the right, and passes beneath the card with the circular aperture. The ends of the moving lines are now obscured, and only the downward motion of the lines is seen in the aperture.

as derived from velocity (flow) fields in conjunction with probabalistic (Bayesian) constraints. The models, known as "energy" models," are built upon arrays of elementary spatiotemporal filters, and such filters, because of their limited receptive fields, are subject to the *aperture problem* (Figure 2.4).[20] As such, the estimate of velocities is inherently *uncertain*, forcing a probabilistic approach.[21]

One of the fundamental probabilistic constraints used by Weiss, Simoncelli and Adelson (*Nature*, 2002) is "motion is slow and smooth."[22] Such a constraint can be applied mathematically, and the model explains a very large array of "illusions." In fact, due to this inherent measurement uncertainty, *all* perception, "veridical" or otherwise, the authors argue, must be viewed as an *optimal percept* based upon the best available information. Applied to the velocity fields defining a narrow rotating ellipse (Figure 2.5), for example, the violation of this "slow and smooth" constraint ends in specifying a non-rigid object if the motion is too fast (Mussati's illusion).[23]

If we were to consider a rotating "Gibsonian" cube, this form becomes a partitioned set of these velocity fields. As each side rotates into view, an expanding flow field is defined. As the side rotates out of view, a contracting flow field is defined. The top of the cube is a radial flow field. The "edges" and "vertices" (i.e., "features") of this cube are now simply sharp discontinuities in these flows. The

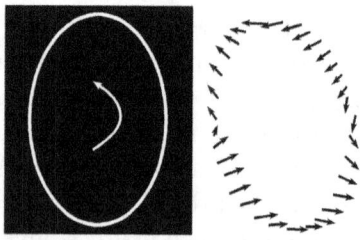

Figure 2.5. The normal velocity components (right) of the edge of a rotating ellipse (left). These tend to induce non-rigid motion.[24]

25

implications of this are concretely displayed in a demonstration discussed by Shaw and McIntyre with a rotating wire cube (Figure 2.6).[25] A cube naturally has a symmetry period of four – it is carried into itself every 90 degree rotation, or four times per a full 360 degree rotation, which is to say its form is invariant (unchanged) after each 90 degree turn. (Symmetry, it should be understood, *is* invariance). When the cube is strobed in phase with or at integral multiple of its symmetry period, for example 4 or 8 or 12 times per revolution, it appears, indeed, as a cube in rotation. But when it is strobed out-of-phase (say, 9 or 11 times per revolution), it becomes a distorted, wobbly, plastic-like or non-rigid object. In this wobbly "not-cube" case, the constraint (invariance) likely being violated via the arrhythmic strobe is this: a regular form displays a regular periodicity in time.

The strobe is essentially taking snapshots of the cube. Yet these snapshots are not sufficient to specify the rigid cubical form we would expect; they are not sufficient to specify the straight lines, straight edges, corners or vertices – the standard static, geometric "features" of a cube. As Gibson long argued, the concepts of our Euclidean geometry – straight lines, curves, vertices, sets or families of forms related by geometrical transformations – while elegant, have little

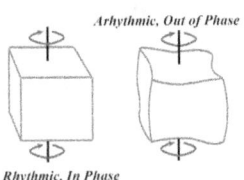

Figure 2.6. Rotating cubes, strobed in phase with, or out of phase with, the symmetry period.

meaning to the brain, i.e., they are not the elements by which the brain constructs a world.[26] Rather, the forms being specified are functions of the application of constraints on flowing fields. The structure of the forms reflects invariants existing over these time-extended flows.

There is nothing static in the ever-transforming material field. The "edges," "vertices" or "surfaces" of a rotating cube do not exist in an instant. Nor its color. There are no "instants." There is nothing static. The brain, simply itself a part of the ever transforming flux, cannot use in its computations that which for it does not exist. Even form can only be derived by imposing constraints (invariance laws) over ever flowing fields. In essence, even the most veridical of forms is simultaneously an "illusion," but yet the best partition of the transforming field the brain can offer.

"Form is only a snapshot of a transition," said Bergson. [27] The eyes are continually in motion. Objects eventually disappear when, in experiments, the position of the object is fixed relative to retinal motion. The brain is at a loss in a static world. The brain is, and is embedded in, an ever flowing material field; it is tuned to this fundamental aspect of reality, and form is obtained by the application of constraints across these flow fields – information inherently uncertain due to the non-fixity.

The misconception of "static" form, derived from the classic metaphysic and Galileo's mis-assignment of form to the "quantitative," underlies the qualia debate participants' failure to grasp that the issue being addressed is the problem of the origin of the *image* of the external field. All seem to think that the origin of the image of the *forms* of the external world is no problem – these are easily "computable" and hence the image itself is no problem, only its "qualities." They fail to grasp that the origin of the image of the forms in the field and of the objects in the field is just as much a problem as the (other) "qualities" of the field – the "rednesses," the "velvets," etc., etc. *None* of these is simply "computable." The statement of the hard problem solely in terms of the origin of qualia has been very misleading. *It is the origin of our image of this field, any image, that is the problem.*

The brain is integrally a part of the abstract continuum of the classic metaphysic. Therefore, when light rays strike objects termed eyes in brain, the abstract, homogeneous motions of the external matter-field, all reducible in time-extent to mathematical points, simply continue in the portion of the field called the "brain." Nowhere in the brain, taken as part of the abstract continuum, can there be anything but more homogeneous points/instants. There can be no actual time-extent of motions through the nerves, no "continuity of time-extended neural processes" – the logical time extent of any neural process is never more than a mathematical point, then another, then another. However one views these motions within the brain, e.g., as maintaining some structural correspondence or isomorphism relative to the always past transformations in the external field or as the processing of invariants in this structure of field motions relative to the body's action systems, it changes nothing. Within the brain, taken as a part of this abstract, homogenous continuum, we can never derive qualities, whether qualities of objects (colors, smells) or of time-extended motions (ignoring that the "object" *is* a motion). We cannot explain how we see a cube "rotating" let alone a "blue" cube. Therefore, all qualia are logically forced, within this metaphysic, into the non-physical, or the mental, or somewhere, anywhere but the abstract continuum. But the step by which this generation of events unto and into another realm can occur, *within the confines of the metaphysic,* remains a dilemma. The structure of the metaphysic makes the step impossible, while leaving the nature of realms outside the structure – e.g., the "mental" – forever incapable of definition or of use to the science that currently operates precisely within this metaphysic.

Bergson on Perception

Bergson's "temporal metaphysic" is equally important to both physics and psychology. For psychology, it provides a very different framework for approaching the hard problem. In this temporal metaphysic, the indivisible or non-differentiable motion of the material field forms an elementary property of *memory* in the field's motion – each (now past) "instant" does not cease to exist as the next (the present) instant appears. It is this "primary memory" – an attribute of the time-evolution of the material field – that supports our perception of "stirring" spoons, "twisting" leaves, "rotating" cubes. Quality is now inherent in this motion of the material field. At the null scale of time, the field is near the homogeneity envisioned by the classic metaphysic, but at ever larger scales of time where the oscillations of the field (e.g., the 400 billion/sec oscillations of the field as a "red" light wave) are "compressed" in the experience or glance of a moment, we obtain ever differentiating quality.

Bergson realized in 1896 that this field is holographic – the state of each "point" in the field is the reflection of, or carries information for, the whole. Noting that there is no "photograph" of the external field developed in the brain, he stated:

> But is it not obvious that the photograph, if photograph there be, is already taken, already developed in the very heart of things and at all points in space. No metaphysics, no physics can escape this conclusion. Build up the universe with atoms: Each of them is subject to the action, variable in quantity and quality according to the distance, exerted on it by all material atoms. Bring in Faraday's centers of force: The lines of force emitted in every direction from every center bring to bear upon each the influence of the whole material world. Call up the Leibnizian monads: Each is the mirror of the universe. [28]

But, as opposed to Pribram, the brain is not simply a "hologram."[29] Rather, to place Bergson's view in modern terms, the brain is the modulated *reconstructive wave* "passing thru" the external, holographic matter-field (Figure 2.7). This brain-embodied reconstructive wave is specifying, or is specific to, always, an image of the *past* motion of the material field – a buzzing fly, a rotating cube. The fly's wing-beats being specified have long gone into the "past," but the indivisible motion of the field – where each present "instant" does not cease to exist – supports this past-specification. The image is right where it says it is – in the field. It *is* the field – the past of the field – at a specific scale of time. The brain dynamics supporting the specification determines this scale of time. The chemical velocities underlying these dynamics are responsible for this. Begin increasing these velocities (equivalently, the energy state) significantly – the fly transitions, from a buzzing fly, to a fly barely flapping his wings like a heron, to a motionless being, to a vibrating, crystalline structure, and on. Again, scale implies quality. We have specification of a qualitative field at a scale of time.

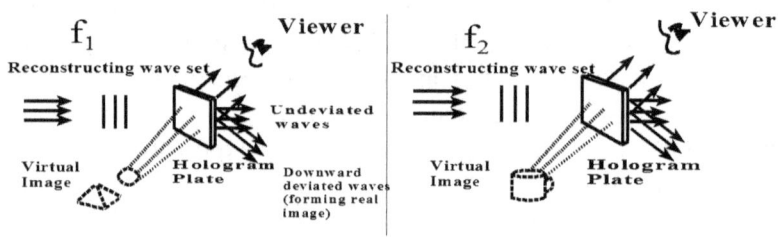

Figure 2.7. Holographic reconstruction. The reconstructive wave, modulated to frequency 1, reconstructs the stored wave front (image) of a pyramid/ball. The reconstructive wave, modulated to frequency 2, now reconstructs the wave front of the cup.

Gibson had come to the notion that the brain must be viewed as "resonating" to the invariance information in events. Resonance is a time-extended phenomenon, and he had come to this because he realized that an invariant that is defined only over a continuous flow of time, e.g., the form of the cube over its flowing fields, cannot be transmitted as a "bit" of information travelling along the nerves. He had seen the invariance information in his texture gradients and flow fields as being incorporated into this resonating brain, causing the brain to be "specific to" the external environment, e.g., specific to a gravel surface receding into the distance. He never explained how an image of the external world arises via this being "specific to" the environment. For this we need Bergson's holographic framework, but the invariance laws defining environmental events is critical, for the continuous modulation of the brain (as a "resonant" wave) is driven by the invariance structure of the external events – the velocity flows defined over the sides of the cube as it is rotating conjoined with its recurring symmetry period, or the motion of the flow field streaming by as we drive.

Due to the continuous motion of the field, this information is always inherently uncertain – we have always an optimal specification of the past motion of the field. In holography, a reconstructive wave, passing through a hologram and successively modulated to different frequencies, successively *selects* information from the multiple, superimposed wave fronts originally recorded on the hologram, and successively specifies each – a pyramid/ball, a cup, a truck, etc. If modulated to a non-coherent (non-unique or composite) frequency, it specifies a fuzzed superposition of the three. There is no "veridical" or God's eye view selection. So too, the brain as a reconstructive wave is selecting information from the transforming matter-field, where the principle of selection is based on information (invariance) relatable to the body's action systems – hence the intimate feedback to and from its motor areas to the visiual areas. In Bergson's succinct phrase, *perception is virtual action*. Following hard on the passage describing the "photograph developed in the very heart of things and at all points in space," he noted:

> Only if when we consider any other given place in the universe we can regard the action of all matter as passing through it without resistance and without loss, and the photograph of the whole as translucent: Here there is wanting behind the plate the black screen on which the image could be shown. Our "zones of indetermination" [organisms] play in some sort the part of that screen. They add nothing to what is there; they effect merely this: *That the real action passes through, the virtual action remains.*[30]

The heron-like fly slowly flapping his wings would also be a specification of the action possible to the body at this new scale of time, in this case, modulating the hand to leisurely catch the fly by the wing.

Given the holographic properties of the field, where the state of each "point/event" reflects the mass of influences from the whole, simultaneously therefore a state of a very elemental "awareness" of the whole, and given the field's indivisible motion defining a primary memory, there is implied, at the null scale of time, an elementary form of awareness defined throughout the field. This is a field property. It is not elementary "constituents" with ad hoc intrinsic and extrinsic properties that must be "composed." This is the old metaphysic, spawned from perception's derivation of "objects" and "motions," still speaking. The specification, then, is simultaneously to a time-scale specific form of this vast, taut "web" of awareness at the null scale. At the null scale, there is no difference between subject and object. Run the scaling transformation in reverse. The fly transitions – initially waves in the field undifferentiated from the perceiving subject, it becomes a crystalline, vibrating being, then becomes the motionless fly, then the heron-like fly slowly flapping his wings, then the buzzing fly of normal scale. Subject is differentiating from object. This is the meaning of Bergson's statement: *"Questions relating to subject and object, to their distinction and their union, must be put in terms of time rather than of space."*[31]

The body/brain as a modulated reconstructive wave passing through a holographic universal field, specifying (or specific to) an image of the past motion of the field's non-differentiable motion, and reflective of possible action at a particular scale of time – this is the elegant solution of the universe to the problem of specifying an image of the external world for its living organisms. Nearly fifty years before Gabor, this was Bergson's insight.

Perception as Virtual Action

For Bergson, the perceived world is the reflection of the possibilities of bodily action. Again, succinctly, perception is *virtual action*. As noted, the fly buzzing by, his wings a-blur, is an index of the possibility of the body's action. Were the fly flapping his wings slowly, like a heron, this would be an index of a yet different possibility, in this case, reaching out slowly and grasping the fly by the wing tip. Note that in each case, this index is simultaneously reflective of a *scale of time*, also a feature of our perception.

That perception is indeed virtual action is indicated by our modern understanding of the processing areas of the brain with their reentrant connections. For example (simplifying greatly), visual area V1, which initially receives the retinal signals, projects to V4 (which handles simple form processing) and V5 (motion processing). Simultaneously V4 and V5 project diffusely back to V1, modulating V1's processing. While the visual areas project to the motor areas, simultaneously the motor areas feedback to the visual areas, modulating visual processing. In fact, counter intuitively unless one is considering virtual action, if we simply sever the connective tracts between the visual areas and the motor areas, the subjects (monkeys in these experiments) go blind.[32]

But supporting this resonating feedback in the neural architecture, there are underlying chemical velocities. It is the base rate of these chemical velocities that determines our normal scale of time, e.g., the world of normally "buzzing" flies. Chemical velocities are subject to modification by catalysts. Were a catalyst (or catalysts) of sufficient strength introduced into the systems underlying the computation and preparation of action, increasing the velocity of chemical processes, then we could expect that the time scale of perception would change. In principle, catalysts of sufficient strength would now allow the system to specify a heron-like fly, barely flapping his wings. By the principle of virtual action, this view of the fly is precisely a specification of how the body can act.

It should be noted that at each scale of time, it is yet invariance laws that specify the event and the possibility of action. Interestingly, as we saw, invariance laws are the only realities in Einstein's relativistic, changing space-time partitions, i.e., the same law, $d = vt$ or $d' = vt'$ holds in each observer's partition. This is but a reflection of perception. Consider a "slow" event in our normal scale, namely the aging of the facial profile. Pittenger and Shaw discovered that this event is specified by a cardioid figure placed over the skull with the aging transformation being a reflection of a strain transformation on the cardioid (Figure 2.8).[33] Were the

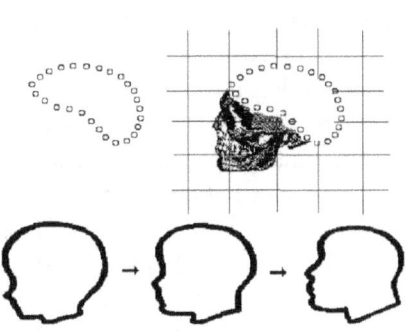

Figure 2.8. Aging of the facial profile. A cardioid is fitted to the skull and a strain transformation is applied. (Strain is equivalent to the stretching of the meshes of a coordinate system in all directions.) Shown are a few in the sequence of profiles generated.

underlying neuro-chemical processes greatly slowed, such that for such a being the facial profile was rapidly changing – now a "fast" event – yet the same invariance information is relating to the action systems, specifying this event and the new possibility of action, e.g., modulating the hand to grasp the rapidly transforming head. As changes in underlying chemical velocities are indeed possible in principle – even raising the temperature can increase these velocities – we must assume that nature has allowed for it and thus must rely on invariance laws. This is the deeper significance of Gibson's insistence on invariance laws specifying perceived events. Indeed it can be argued that this holds for the high-metabolism chameleon flicking the grasshopper off the leaf, the frog snatching the fly, the chipmunk scurrying by – all employing invariance laws in their unique space-time partitions.

Special Relativity and Perception

The change of scale and form for the fly – say, from buzzing to heron-like – is not merely "subjective," or, in philosophical phrasing, "a subjective modification" of experience. This is an objective effect. Virtual action, straightforwardly, makes a prediction on action relative to the increase or decrease of the velocity of underlying processes. In principle, this is a testable consequence albeit difficult today. The question is, does Special Relativity also make a prediction, and if so, what?

Let us consider the case of two observers, X and Y. We take the X system to be stationary, and Y moving relative to X at high uniform velocity. Assume there is a fly in X's system. X, at his normal velocity of processes, i.e., at his time-scale, perceives the fly as a blur. The fly, which X is observing, travels one of X's distance units using sixty wing-beats. It does this in one of X's time units, say a second. Y, moving at great velocity, has much expanded time units (and contracted space units), the time units increasing as he moves nearer to the speed of light. However, this is as X computes these units relative to his stationary system. The complimentary case is Y's (in motion) view of the space-time of X. The Minkowski diagram (Figure 2.9) shows this situation. The rhombus OFGH is gradually collapsing like a scissors as the velocity of Y increases. The tangent to the hyperbola, GF, drops lower and lower below X's time unit, displaying that the time units of X, as Y sees them, are contracting steadily. Eddington had us imagine

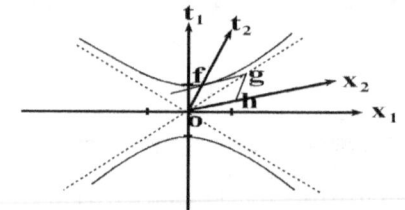

Figure 2.9. The Minkowski diagram.

that at O, X lights up a cigar that lies along x_1 and has a very longish length of one space unit. The cigar burns one of X's units of time, being represented by the line t_1 and extending to its first unit. Y would now see the cigar as burning longer for X, in fact, as the tangent drops as v increases, it would last many units of X as assigned by Y. This could equally be X himself, aging (a form of "burning") many more time units than Y. Simultaneously, the space units of X, as Y sees them, are increasing. Thus note that GH would fall outside the space unit of X – the cigar is longer.[34]

Now it might be said that the fly, flying the length of the cigar lying along x_1, is flying a longer distance as far as Y is concerned since he determines X's space units have expanded. But the distance that the fly traverses in sixty wing-beats – however great or small the distance is *measured* to be – this distance holds a fundamental "causal flow" or invariant that relativity and its measurement procedures cannot alter. If we mark this distance by two markers, A and B, the fly will buzz from A to B in sixty wing-beats, no matter what the reference system from which he happens to be viewed. It is the "sixty wing-beat distance invariant."

We start from this. The fly flies this distance every day, from the cereal bowl to the sticky spoon on X's table, in sixty wing beats. Relativity, simply because Y goes into motion, contains no inherent justification for altering this.

Assume that the rocket is moving at 80% the speed of light. Given Y's view of X as having contracted time units, the same sixty wing beats require 1.66 seconds as assigned to X by Y. So, now we partition this sixty wing beats (an invariant causal flow) across the 1.66 seconds. In X's normal system, at sixty wing beats/second, there are six wing beats in each $1/10^{th}$ second, and X can normally perceive or discriminate one wing-beat per $1/10^{th}$ second. Thus at six beats per each $1/10^{th}$ second, he sees a blur. In the new partition assigned by Y, with sixty beats partitioned over the 1.66 seconds, X sees only 3.6 wing beats in each $1/10^{th}$ second. It is, shall we say, less a blur. The fly appears to be buzzing more slowly. X's time (his perception of the rate of events) is slower, despite the fact that his velocity of processes has not changed. This is clearly absurd, yet this is exactly what is required of the world of X if we ignore reciprocity, and if these transformations are ontological enough to support Y's eventual return as more youthful than X.

On the other hand, there is the effect on Y, whose time units are expanded and space units contracted. In Y's moving system, a fly is buzzing across the table in the rocket cabin, again using sixty wing beats from A to B. It requires only .6 of the expanded Y-second for the distance to be covered. The invariant sixty wing beats are partitioned across this amount, therefore becoming ten beats per each $1/10^{th}$ second, and thus the fly is now more of a blur, despite the unchanged velocity of processes. It can be argued, just as Eddington notes, that due to the rocket's velocity, Y's processes are retarded. But in fact everything in Y's reference system is retarded, to include the fly and its buzzing from A to B. In effect, we have simply subtracted a constant across all motion values of the system, and the problematic modification of perception just noted still holds. In essence, psychology contradicts physics.

In this analysis, I have stayed consistently within the implications displayed in the Minkowski diagram, that is to say, within the case where Y is consistently the one in motion, X stationary. If we want to set X in motion, we need another diagram, and the situation simply reverses.

What is wrong here? We must return to examine the inherent reciprocity in relativity.

Chapter II: End Notes and References

1. Bergson, H. (1896). *Matter and Memory,* p. 255.
2. Smythies, J. (2003). Space, time and consciousness. *Journal of Consciousness Studies, 10,* 47-56.
3. Weyl, Hermann (1922/1952) *Space, Time, Matter.* Dover, p. 217 (emphasis added).
4. Eddington, A. (1966). *Space, Time and Gravitation.* Cambridge: MIT Press, p. 57.
5. Dunne, J. W. (1927). *An Experiment with Time.* London: Faber and Faber.
6. Smart, J. J. C. (1967) Time. *The Encyclopedia of Philosophy.* New York: Collier-MacMillan.
7. Smythies, J. (2003). Replies from John Smythies. http://tech.groups.yahoo.com/group/jcs-online/message/2582.
8. Robbins, S. E. (2002). Semantics, experience and time. *Cognitive Systems Research, 3,* 301-337.
    Robbins, S. E. (2007). Time, form and the limits of qualia. *Journal of Mind and Behavior, 28,* 1-25.
9. There is a mythology that these paradoxes have been resolved by Russell (1903) and/or modern mathematics. While Bergson showed that all four paradoxes have exactly the same root cause in an abstract space, Russell, having missed the point, actually accepted the fourth paradox ("a duration is the double of itself") as a physical reality. The mathematical "resolutions" are inherently limited to a spatial treatment and, in "taking a limit," simultaneously invoke hand waving over infinity in the operation (cf. Bergson, 1907/1944, pp. 335-340). For Achilles' pursuit of the Tortoise, the paradox is considered to have a solution wherein both the distance traversed and the time taken to do so are treated as an infinite geometric series with each successive term multiplied by 1/2, each term for distance and time thus growing successively smaller. Achilles traverses an infinite number of distance intervals and similarly an infinite number of time intervals before reaching his Tortoise-goal. Such a series converges, and by taking a limit as n (the number of terms) approaches infinity, we arrive at a finite answer for the time required. The motion has thus been structured as an infinite number of distance and time intervals, with the solution relying on a mathematical technique that conveniently gets rid of the infinities, i.e., it simply ignores that infinity *means* infinity while leaving the problem of Achilles' motion within the static instants, or for that matter, the actual *physical* resolution of the infinities, unsolved.
    In other words, there is no physical model of how either Achilles or the motion of the universal field of which he in an intrinsic part could resolve this, and unfortunately, when we are constructing a model of the brain, it is achieving a physical reality that is all important. Further, for each of these successive intervals, the scheme assumes that the velocity of Achilles in fact is *fixed* – determined. But the values of each successive term are not actually representative of the times at which Achilles is in a particular position, but rather of the fact that he is *passing through* the interval, however tiny it might be. The associated time (t) values and spatial (d) values, and thus the very derivation of velocity, is again reliant on the imposed static backdrop, a backdrop in which Achilles actually is passing through, but to which he has no intrinsic, or yes, *physical* relation. Whether we take a time value of 1 sec or 1/1000 sec (or however small), we

have only an interval in which Achilles is passing through a certain distance interval (e.g., 1 m).

10. Bergson, H. (1896). *Matter and Memory,* p. 255.

11. Bergson, H. (1896). *Matter and Memory,* p. 259.

12. Bergson, H. (1896). *Matter and Memory,* p. 259.

13. Bergson (1889). Time and Free Will, p. 111.

14. Lynds, P. (2003). Time and Classical and Quantum mechanics: Indeterminacy versus discontinuity. *Foundations of Physics Letters 16,* 343-355.

15. Nottale, L. (1996). Scale relativity and fractal space-time: applications to quantum physics, cosmology and chaotic systems. *Chaos, Solitons and Fractals, 7,* 877-938.
Feynman, R. P. & Hibbs, A. R. (1965). *Quantum Mechanics and Path Integrals.* New York: MacGraw-Hill.

16. Chalmers, D. (1995). Facing up to the problem of consciousness. *Journal of Consciousness Studies,* 2(3), 200-219.

17. Robbins, S.E. (2004). On time, memory and dynamic form. *Consciousness and Cognition, 13,* 762-788.

18. Manzotti, R. (2008). A process-oriented view of qualia. In E. Wright (Ed.), *The Case for Qualia.* Cambridge, Massachusetts: MIT Press, pp. 175-190.

19. Gibson, J. J. (1950). *The Perception of the Visual World.* Boston: Houghton-Mifflin.
Gibson, J. J. (1966). *The Senses Considered as Visual Systems.* Boston: Houghton-Mifflin.

20. Adel son, E., & Bergen, J. (1985). Spatiotemporal energy model of the perception of motion. *Journal of the Optical Society of America,* 2 (2), 284-299.
Watson, A. B. and Ahumada, A. J. (1985). Model of human visual-motion sensing. *J. Opt. Soc. Am. A.,* 2, 322-341.

21. See for a review, Robbins, 2004, op. cit.

22. Weiss, Y., Simoncelli, E., & Adelson, E. (2002). Motion illusions as optimal percepts. *Nature Neuroscience, 5,* 598-604.

23. Mussati, C. L. (1924). Sui fenomeni stereocinetici. *Archivo Italiano di Psycologia,* 3, 105- 120.

24. Weiss, Y., and Adelson, E. (1998). Slow and smooth: a Bayesian theory for the combination of local motion signals in human vision. MIT A. I. Memo No. 1624.

25. Shaw, R.E., & McIntyre, M. (1974). The algoristic foundations of cognitive psychology. In D. Palermo & W. Weimer (Eds.), *Cognition and the Symbolic Processes,* New Jersey: Lawrence Erlbaum Associates.

26. The fact that we find cells that are sensitive to the motion of straight lines in the small area of their receptive fields does not mean that the brain works, or is programmed, as though space were Euclidean, i.e., that this construct – itself a derivative of perception and human conceptual development – is inherent in or built into the brain. In fact, these elements are no longer considered the building blocks of a scene (Nakayama, 1998).
Nakayama, K. (1998). Vision fin de siècle: A reductionistic explanation of perception for the 21st century? In J. Hochberg (Ed.), *Perception and Cognition at Century's End.* New York: Academic Press.

27. Bergson, H. (1907/1944). *Creative Evolution,* p. 328.

28. Bergson, H. (1896/1912). *Matter and Memory.* New York: Macmillan, p. 31.

29. Pribram, K. (1971). *Languages of the Brain.* New Jersey: Prentice-Hall.
30. Bergson, 1896, op. cit., p. 31-32, emphasis added.
31. Bergson, 1896, op. cit., p. 77, emphasis added.
32. Weiskrantz, L. (1997). *Consciousness Lost and Found.* New York: Oxford.
33. Pittenger, J. B., & Shaw, R. E. (1975). Aging faces as viscal elastic events: Implications for a theory of non rigid shape perception. *Journal of Experimental Psychology: Human Perception and Performance* 1: 374-382.
34. Eddington, A. (1966). *Space, Time and Gravitation.* Cambridge: MIT Press, p. 57.

CHAPTER III

Half-Relativity, the GTR and Other Problems

For the geometer all movement is relative: which signifies only, in
our view, that none of the mathematical symbols can express the
fact that it is the moving body in motion rather than the axes or
points to which it is referred.

— Bergson, *Matter and Memory*, p. 256

The Role of Reciprocity

What has gone wrong when we view STR's treatment of the fly? There is the strange picture of Y's view of X's altered perception of events in X's own system. But let us ignore this. One aspect of the problem is more elementary. As noted, when we represent the situation of X and Y in the Minkowski diagram, we have fixed on one observer, X, and set all other systems in motion relative to him. The Minkowski schema represents the adjustments in time and space units necessary to preserve light-velocity invariance for all other systems. But it cannot represent reciprocity. We could equally have fixed on Y and set all other systems in motion with respect to *him*. This, again, requires another diagram, and so on for each observer upon whom we fix.

Given this, we must ask the fundamental question: is the effect on either X or Y a real effect? Y, we know, could equally declare his system to be at rest, and X in motion relative to him. Clearly, the effects cannot be real from this perspective. The different "times" and "distances" represent only the observer's method of keeping his measurements consistent with light-velocity invariance. STR, from this perspective, fails to justify, either for X or for Y, a different perception of the fly based on the observer's motion. If we respect the inherent reciprocity of reference systems in STR, there is no contradiction with the relativity of perception. STR is at worst neutral with respect to a causal flow in time (the fly) invariant to both X and Y. Only if we insist that STR implies a real effect is there a contradiction.

It must be clearly understood again here that I am not denying the empirical facts, e.g., increase of life spans in mesons, or the retarded clock carried by the jet, or increases in mass. The empirical evidence is not in dispute. These are real effects. What is in dispute is the use of STR to explain the empirical evidence; it is used inappropriately in attempting to do so. The structure of reciprocity intrinsic to STR is being ignored.

Half-Relativity

"Half-relativity" is what Bergson termed the asymmetric use of STR.[1] The Lorentz equations are applied to the meson; the life span increase falls out via t'. End of explanation. As noted already, A. P. French, in a textbook that attempts to maintain clarity, in a section entitled "Relativity is Truly Relative," flatly states that the time dilation (just as the length contraction of the Michelson-Morley apparatus) as observed for a meson is not a property of matter but something inherent in the measurement process.[2] He goes to the rare extent of actually showing *two*

Minkowski diagrams, one for each observer (as though there were a small observer on the meson), to show the symmetry of the changes in *each* system. Just as Bergson argued earlier, French notes that were an observer to compute t' as the meson falls to the earth, the tiny observer on the meson is equally allowed to say that he is stationary and the earth moving towards the meson. This is to say we have here, in French's terms, a "measurement effect." Thus, when French treats the twin paradox, he invokes the *asymmetry* introduced when the twin on the rocket turns around to return, therefore introducing a new inertial frame .[3] STR is used to compute the different (shorter) "time" of the traveling twin for each leg of the trip, thus ascribing the magnitude of the difference to v. But he assumes, in conjunction with this, that it is the asymmetry introduced by the turn-around that is required to support the real (aging) effect, i.e., as *a real property of matter*. Clearly, if one twin is now gray and has a long beard, we have a change that is a real property of matter. Thus he argues that STR, factoring in this asymmetry associated with the turn-around and its acceleration, and due to the fact that a time difference value can be derived due to v, can indeed handle the twin paradox. Yet he has earlier painstakingly built the case, to the point of doubled Minkowski diagrams, that the structure of STR demands symmetry (reciprocity), and given this symmetry, it does *not* explain any changes as real properties of matter.[4] In essence, the entire explanatory burden for aging as a real effect now falls on the asymmetry introduced by the change in inertial frame. *But where is this theory!?? That is, where is the theoretical framework supporting how and to what magnitude introducing an asymmetry affects the physiological processes underlying aging?* Or why the asymmetry can be introduced into STR? More precisely, where is the theory that explains how introducing an asymmetry now allows the use of the Lorentz equations independent of, or outside of, the symmetric, reciprocal structure provided in STR?

In the comparison between X and Y above, we only asked Y to be in uniform relative motion at velocity v, just as in the meson case, just as in the Michelson-Morley case. This comparison could care less about Y's return or differential accelerations. We don't need a rocket. While X sits by the kitchen table watching the fly, Y could travel by on his tricycle, and the same relativistic laws hold.[5] Nevertheless, there are those that would simply classify this case as the twin-paradox, invoke the existence of accelerations, and move the problem and the effects involved into the General Theory. All of the effect can then be assigned to acceleration(s). This reaction is extremely problematic. If we seize upon any accelerating component of a motion (which one can always find, even for the startup of the tricycle) to allow us to get to the safety of the GTR, then what if anything is the province of STR? The physics would be in danger of becoming a shell game, shuffling an explanatory pea between STR and GTR. If we are doing this to avoid reciprocity, then the argument that STR, with its inherent reciprocity, fails to explain any of these effects is effectively conceded, and this lynchpin in its being a theory of time – its ability to explain these effects – is removed.[5] Note again, it is not the aging effect, it is *all* asymmetric effects – jet carried clocks or long living mesons – that would have to be so moved into GTR for consistency. One dismisses the above comparison of X and Y into the GTR then only with difficult consequences.

Thus others (as well as French) have argued, as Eddington appeared to believe, that the twin-effect is perfectly consonant with STR. But to stay fully within the context of the Special Theory without bringing in gravitational field changes, Salmon envisaged a rocket ship (A) departing earth and passing another (B) coming in the opposite direction at the same velocity. At the point of meeting, the two exchanged signals to coordinate their clocks.[6] B continued on to earth where clocks were compared, and of course, in a triumph for the theory, an earthbound observer's clock showed a greater passage of time than B's. This appears to be ironclad, yet there is a problem. Reciprocity has not been avoided. The observer in A takes with him his own reference system. Since no reference system is privileged, he has equal right to declare himself at rest and everything else in motion relative to him, including the earth, the earthbound observer, and the earthbound observer's clock. When B passes A and signals are exchanged, will they then reflect a decrease in the rate of A's time? Hardly, given A is at rest. Only the author of the argument happens to believe A is in motion, but he forgot to ask A.

Davies (1977) resolves the twin paradox by flatly assigning the aging differential to the turn around at the target star and the homeward acceleration of the rocket (pp. 43-44). Yet, like French, he applies the Lorentz equations, claiming that he has also preserved the symmetry, a fact his table of durations (p. 44) obviously belies, for only the rocket clock shows a consistent, time-expanded 4.8 light years for each leg – the rocket is clearly the only object moving to Davies. Davies (1995) drops the clear emphasis on acceleration as the root cause of the aging. He does declare "there is no paradox" because the symmetry is broken due to accelerations in the necessary stop and return of the rocket, but never mentions this again. Ignoring the consequent inapplicability of STR, he again proceeds to apply the Lorentz transformations (with what justification?). In essence, he notes that at 80% of the speed of light, earthbound twin Ann would see the clock of the rocket-twin (Betty) as running .6 of earth-Ann's. Symmetrically, rocket-Betty, viewing herself as stationary, sees earth-Ann's clock as running .6 of Betty's. This symmetry holds for each leg – the outward and the homeward bound. In Davies' scenario, it is rocket-Betty who returns having aged less, not earth-Ann, and he claims that he has resolved Dingle's (1972) critique that in this case, "each clock runs slower relative to the other," in other words, a critique which says precisely that there can be no ontological status here.[7] Given the symmetry he took great pains to describe, Davies conveniently never tells us why earth-Ann does not also have the distinction of aging less.[8]

The twin-paradox is disturbing precisely because it epitomizes, very concretely, the inconsistency relative to standard use of STR. It highlights a very real effect, e.g., a youthful man versus a hoary old one, that cannot simply be assigned to a measurement process. Interestingly, Einstein himself, in a (little known) 1918 article, attempted to preserve reciprocity and the asymmetrical effects together by arguing that indeed the rocket ship could be considered stationary, its motors only neutralizing the pull of the earth as the earth recedes.[9] (This is what I meant re Einstein's "yes and no" position on Langevin – he is at least trying here to reserve reciprocity.) But he then argued that it would require such tremendous field changes to move the earth and bring it back that the earth twin would undergo rapid aging. The reciprocity and the paradox denying the reciprocity appear resolved (just as French argued). But now, ignoring the ad hoc, physically unrealizable fields, it is not clear of what use relativity is here at all. Its mathematics, with its intrinsic reciprocity, now does not accurately describe the phenomenon – we can clearly distinguish the two systems via gravitational effects – and it would seem logically prior to have a theory relating gravitational changes to a model of the physiological processes driving aging – this in itself being sufficient to account for the phenomenon without appealing to changes of "time" itself. The one-way application of the Lorentz transformations would then appear in retrospect to be but a convenient empirical description of these events, but a deeper theory would provide a model of the processes involved (as Lorentz himself attempted).

The Half-relativity of 1905

Einstein, for all practical purposes, began assigning real effects due simply to v, ignoring reciprocity, in the very paper that introduced the theory, in 1905. In the paper, he quickly invokes the reciprocity implied in the first postulate, having us envisage a rigid sphere of radius R, at rest in the moving system.[10] At rest relative to the *moving* system, he notes, it is a sphere. Viewed from the "stationary" observer the equation of the

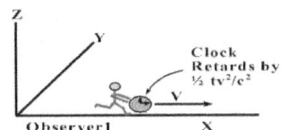

Figure 3.1. The observer causes the clock to retard simply by moving it at velocity v within his own reference system.

sphere's surface gives it the form of an ellipsoid, with the X dimension shortened by the ratio $1:(1 - v^2/c^2)^{1/2}$. He notes (the reciprocity) then immediately: "It is clear that the same results hold good of bodies at rest in the 'stationary' system, viewed from a system in uniform motion." Two paragraphs from this point he notes the "peculiar consequence" that were there two synchronous, separated clocks A and B in the stationary system, and if A is moved to B with velocity v in time t, it will lag behind B by ½ tv²/c². The structure of reciprocity is already being voided here – we are dealing only with an effect in the stationary system, not relating the two systems. The observer in the stationary system can simply move the clock from A to B to fulfill Einstein's condition, and the effect is simply ascribed to v (Figure 3.1).

This conclusion is quickly reinforced. Within another paragraph, Einstein, extending this to "curvilinear motion," states flatly that this result implies that a clock at the equator must go more slowly, by a small amount, than one situated at the poles, i.e., again two clocks in the same system, in this case the earth.[11] Physicists accept this equatorial clock retardation naturally as a real effect. The effect had to be factored in to Hafele and Keating's jet-carried clock experiment. Yet reciprocity demands that the clock on the equator be stationary, the observer at the pole spinning around. Now it is not a real effect. This is likely not very tasteful. Yet this conclusion regarding v as already producing real effects in 1905 is doubly reinforced when it is considered that the equator-clock is an exact analogue to Einstein's future thought experiment (introducing GTR) of the rotating disk. Now the observer leaves the center of the disk, moving along a radius to the rim and back, while carrying a clock (Figure 3.2). Upon his return the clock is retarded. The thought experiment used this result as a very real effect. Yet

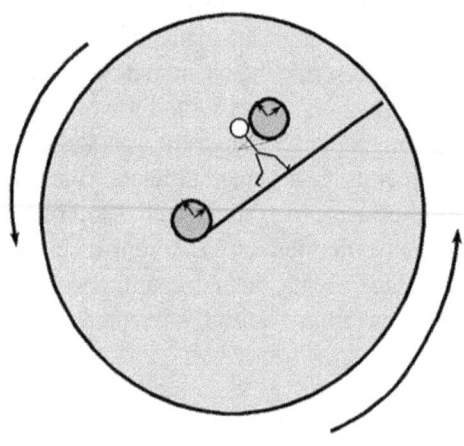

Figure 3.2. The Rotating (therefore accelerating) Disk of GTR. As the observer advances along the radius of the disk, his clock becomes increasingly retarded.

why? The observer takes with him, at every point he occupies, his own proper time. He should return with the clock unchanged.

The "Comfort" of the GTR

It is neither my intention nor scope here to delve deeply into the General Theory. It is invoked so often as an escape valve for these problems that something more should be said that gives the escapees at least a little pause. The issue, further, goes to the heart of the origin of these real effects. The comfort of assigning these real, ontological changes to the GTR arises from the tenet that acceleration breaks the symmetry or reciprocity of systems. Because, as A. P. French did for example, we can invoke acceleration as being involved when the rocket turns around and heads back to earth, we can fix the retarded aging on the acceleration as the cause. Acceleration is considered *real or absolute*; it is not subject to the reciprocity of reference systems as is uniform motion (velocity). That is, it is not the province of STR.

We have seen the problems with the "rotating disk" thought experiment. Let me reprise here an argument that can be found in Bergson's *Duration and Simultaneity* for the reciprocity of accelerating systems as well.[12] Is a "system" then really the unity or non-decomposable entity as it is treated or conceived to be in relativity? Is our body or an automobile for example not in actuality a multiplicity of systems? The auto with our body inside (system B), in uniform motion, is in perfect reciprocal motion to the road (system R). It now accelerates, we feel a "jolt" – the symmetry or reciprocity of B/R we say is broken. The "jolt" signifies that the physical points in our body do not maintain unchanging positions with respect to the car or even to themselves, neither do the points of the automobile. There are in fact multiple systems, B', B", B"', etc. For the (at rest) observer, in R, each of these systems now has its own time, t', t", t"', and it can be argued then that the reciprocity of R to each of these is complete. To see this from the opposite perspective, imagine that B is an infinitesimal point, truly a single, non-decomposable system. This point accelerates, i.e., it merely increases its velocity. Is there the jolt due to the acceleration, i.e., is there any loss of symmetry with respect to R? So this is where the conceptualization of any and all reference systems as a unity could be leading us astray. Symmetry breaking via acceleration may not then be as true a case as one might think. There would still be reciprocity. And then, sending the twin-paradox to GTR solves nothing, but leaves us in the same problem re real effects.

Acceleration was seized upon as an equivalent to gravity by Einstein in his thought experiment on riding in an elevator. In a uniform motion up, I notice nothing, but if the elevator accelerates, I feel a force. I can know I am moving. The motion is not relative, nor subject to STR. This principle sits at the heart of GTR. Acceleration is a real force. One can question Einstein's conclusion. Bergson argued simply that acceleration cannot be distinguished from velocity in the sense relativity claims – velocity is a rate of change in position over time, acceleration simply the rate of change of the rate of change of position. That is, one is the first derivative of change in position with respect to time, the other is simply the second derivative. Why is the second derivative so privileged over the first? Physicist, Ling Jun Wang, refines this argument, deriving the generalized Lorentz equation for t' in the context of acceleration.[13] If we cannot integrate over infinitesimal velocities, he argues, as did Bergson also argue, we have undercut all of physics. Wang's equation completely undercuts any appeal to the GTR due to acceleration in the twin paradox; in fact it implies a question to the foundation of GTR.

The source of GTR's foundation via this principle is in the attempt to use *force* to explain the *real* motion of the matter-field. Simultaneously it represents the inability of the classic metaphysic to come to grips with the primacy of motion. In the classical metaphysic, we have seen that all motion is relative. An object can move from point to point across the continuum of positions, or the continuum (coordinate system) can move beneath the object. But we have also seen that there must be *real* motion in the matter-field – stars explode, trees grow, couch potatoes get fat. How can we distinguish real motion from that which is merely relative and which becomes rest on a change of perspective? "Force" appears as the natural answer. Force is naturally seen, in our classic framework, as that which imparts "motion" to "objects." Real motion emanates from a "force." But this is the problem: force is only a function of mass and velocity, where again, velocity is the *rate of change of position*. Force, or f = ma, is measured by the degree of acceleration it produces in the body (or mass, m). In turn, again, acceleration is only the rate of change of the rate of change of position. We are always dealing, in other words, only with *change of position*. In other words, these movements are still relative. The force, one with these relative movements, does not escape this relativity.[14] Force is no more absolute than the movements; it cannot serve to distinguish real or absolute motion. It is not a "cause." It is, expressed in f = ma, as physics is cautious to treat it, an invariance law. Thus, as Bergson, noted:

> It is in vain, then, that we seek to found the reality of motion on a cause which is distinct from it: analysis always brings us back to motion itself.[14]

As Bergson also argued, if one seeks the principle of absolute motion in force, because of the inescapable relativity of movements in the abstract continuum, you are forced back to the principle of an absolute space with absolute positions, a state for which Einstein's curved space qualifies.

Why is the problem of "real effects" significant? There are three reasons. Firstly, if STR is being used inappropriately as an explanatory device where the one-way use of the mathematics just happens to work, then physics should be searching for the true explanation. It could be extremely instructive, if only for the apparent return of the ether, which formerly housed some of these effects (again, in Lorentz's mind for example), in more sophisticated form as the quantum vacuum. There are probably any number of ways, for example, to account for the life-span increases of mesons without resort to the mystical "changes of time" required by STR, i.e., *by describing a concrete model with actual physics of the changes.* Consider just one.

J. J. Thomson's model of the electron, as just one possible example of an approach, saw the electron as a special case of an electric current. In motion, a current naturally generates a secondary counter-EMF – a resistance to its own motion. So too, as a special case of a current, would a single electron, and as its velocity increases with the accompanying counter-EMF, we reach a singularity, for the resistance to motion of the electron (i.e., its inertia or mass) eventually increases to infinity.[15] Now suppose that a meson is a group of electrons and positrons, say three electrons and two positrons, or five electrons and three positrons, etc., and that a feature of a positron is that it accepts energy from the electron but radiates that energy away into the surrounding field. (A single electron-positron pair simply annihilates itself very quickly.) The time required for the eventual complete radiation away of the group's energy is thus a function of a certain synchrony of this energy acceptance/radiation away established within the group, this being "decay." Now putting such a group in motion will retard this radiation due to the counter-EMF, the decay rate ever decreasing with speed, and increasing the group's lifespan.

I am not saying this is necessarily a correct model; the point is that this is an example of a concrete physical model of such decay and "lifespan" extension as opposed to a mystical "change of time." (See also, Aspden, 1969, 1972; Aspden derives e=mc[2] without the aid of STR).[16] If one examines Bergson's exchange (circa 1924) in *Revue Philosophique* with the physicist Andre Metz on the subject of the twin-paradox and the correct interpretation of relativity, it becomes apparent that, despite Bergson's clear demonstration of the inconsistency in claiming the time changes of STR can be used to explain real effects such as the meson's lifespan extension, the stubborn resistance Metz displayed to accepting this came because he simply could not come to grips with the concept that there might be any other way to explain such phenomena other than by special relativity.[17]

Secondly, there is now the contradiction with the psychology of perception just discussed and which I hope would merit at least some review. Thirdly, if we cling to the idea that STR can explain real, asymmetric effects, then we are equally clinging to the reality of the relativization of simultaneity, i.e., to the *real* breakup of simultaneity into successive moments in time, and vice versa. It is this implication that I wish to further question.

Chapter III: End Notes and References

1. Bergson, H. (1922/1965). *Duration and Simultaneity With Respect To Einstein's Theory.* Indianapolis: Bobbs-Merrill.
2. French, A. P. (1968), p. 114.
3. French, A. P. (1968), pp. 155-156.
4. I have been posed one objection or "solution" to this problem, yes, by a reviewer, stated as follows: "The twin leaving and returning on the rocket ages less because his worldline between departure and return is shorter. And the length of the worldlines is observer in-variant." This is a strange misconception, mis-statement and convolution. The "observer invariance" is only defined within the structure of symmetric (reciprocal) transformations created by both observers. There is no "invariance" with but one observer. But then it is this very symmetry that makes it impossible to use relativity to explain changes as real properties of matter.
5. Brillouin (1970) would argue that a reference system must be very massive to reduce all action-reaction effects. The tricycle, let alone an abstract "coordinate system," would not qualify in his opinion. The same point however can be made with a more massive system going by the table. But I do not believe that Einstein was concerned at all with this dis-tinction, the geometry being the overriding consideration.
   Brillouin, L. (1970). *Relativity Reexamined.* New York: Academic Press.
6. Salmon, W. (1976). Clocks and simultaneity in special relativity, or, which twin has the timex? In P. K. Machamer & R. G. Turnbull (Eds.), *Motion and Time, Space and Matter.* Ohio State University Press.
7. Dingle, H. (1972). *Science at the Crossroads.* London: Martin Brian & O'Keeffe.
8. Davies, P. (1977). *Space and Time in the Modern Universe.* London: Cambridge University Press.
   Davies, P. (1995). *About Time: Einstein's Unfinished Revolution.* New York: Simon & Schuster.
9. A translation of this paper is discussed in Dingle (1972, pp. 191-200).
10. Einstein, A. (1905/1923). On the electrodynamics of moving bodies. In H. A. Lorentz, A. Einstein, H. Minkowski, H. Weyl. *The Principle of Relativity.* New York: Dodd Mead, section 4, p. 48.
11. Einstein, A. (1905/1923). On the electrodynamics of moving bodies. In H. A. Lorentz, A. Einstein, H. Minkowski, H. Weyl. *The Principle of Relativity.* New York: Dodd Mead, p. 50.
12. Bergson, H. (1922/1965). *Duration and Simultaneity*, pp. 173-176.
13. Wang, L. (2003). Space and time of non-inertial systems. *Proceedings of SSGRR 2003*, L'Aquila, Italy.
14. Bergson, H. (1896/1912), op. cit., p. 257.
15. On these concepts and models, see Kessler, J. (1962). *The Energy of Space.* Published by the author.
16. Aspden, H. (1969). *Physics Without Einstein.* London: Sabberton.
    Aspden, H. (1972). *Modern Aether Theory.* London: Sabberton.
17. Gunter, P. A. Y. (1969). *Bergson and the Evolution of Physics.* University of Tennessee Press., pp. 123-135.

CHAPTER IV

The Simultaneity of Flows

In the theory of relativity, the slowing of clocks is
only as real as the shrinkage of objects by distance.

— Bergson, *Duration and Simultaneity*[1]

The theoreticians of relativity never mention any simultaneity but
that of two instants.

— Bergson, *Duration and Simultaneity*

Now the from the simultaneity of flows we would never pass to
that of two instants, if we remained in pure duration for every
duration is thick; real time has no instants.

— Bergson, *Duration and Simultaneity*

The Relativity of Simultaneity

In Figure 4.1 we picture three points, A', B', and C' in Y's moving system placed along the direction of this motion. Each will be a distance L from each other. We will assume Y is at point B', and the system is moving with velocity v. From the viewpoint of the stationary X, these three events are not simultaneous. The clock at A' registers a time slightly behind that of B', while the clock at C' is somewhat ahead. The greater the value of v, the greater this lag and lead time respectively. Both times are given by Lv/c^2 seconds. As v approaches the speed of light c, the maximum difference becomes L/c seconds.

If we drop a perpendicular from A' to K', this line will symbolize all the past events at A'. Since we see that the clock is slow at A', and Y then supposedly looking at past events, this line displays the maximum reach into this past. Likewise the line upwards from C' to H' shows the maximum of the future. Now we can draw yet another line of simultaneity, this one running to (hypothetical) points

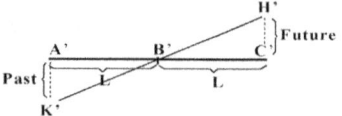

Figure 4.1 Planes of simultaneity

D' (between C' and H') and E' (between A' and K'). Its divergence from the original line A'B'C' is a function of the speed v. Further, were the difference in v between the X and Y systems infinitesimally small, there would be a line barely divergent from A'B'C' representing the fact that at even the most infinitesimal velocities, we see the breakup of simultaneity begin, radiating from the most minute point or distance from B', increasing in degree towards A' and C'. There are any number of such lines.

What is the reality here? Imagine that Y is moving at an infinitesimally small velocity relative to X. For practical purposes, X's line ABC and Y's line A'B'C' are virtually coincident. But yet, even at the most minute velocity, simultaneity has begun to break up at the most infinitesimal point or distance from B, increasing in degree as we approach A' or C'. Now Y moves at a much higher velocity. X now notes the difference in Y's clocks. He is forced to assign events at A' deeper and deeper into Y's past as v increases, and to assign events at C' farther into the future. He does this by the very fact that he needs to keep the velocity of light invariant as per the Lorentz transformations. But Y can equally say he is at rest. He continues to note the simultaneity of events at A', B', and C'. He now notes the same breakup of simultaneity for X. Again the question becomes, is the conversion of simultaneity to succession real? Is it more than a notational convention required for the consistency of measurements between the two systems? Can this possibly be true of the flow of time?

Rakić's Critique

Perhaps we are asking here, though relativity has hitherto withstood the test of time, whether time has truly been used to test STR. Natasa Rakić was able recently to show that there is a very reasonable feature of time that STR must incorporate, yet is not definable in terms of the causality relation inherent in Minkowski space-time.[2]

Consider Figure 4.2. The left side depicts the light-cone concept of relativity. The lines of the X are rays of light. These rays show the path of the fastest possible causal action (light speed) that could emanate, so to speak, from an event. Any thing/event falling within the top "V" of the X can be influenced or causally affected by the event. There is a flip side, or bottom of the X. Anything inside the bottom of the X can influence or causally affect the event. On either side of the X is the "causal elsewhere" of the event. Any happening here cannot affect the event. It would have no causal relation to the event.

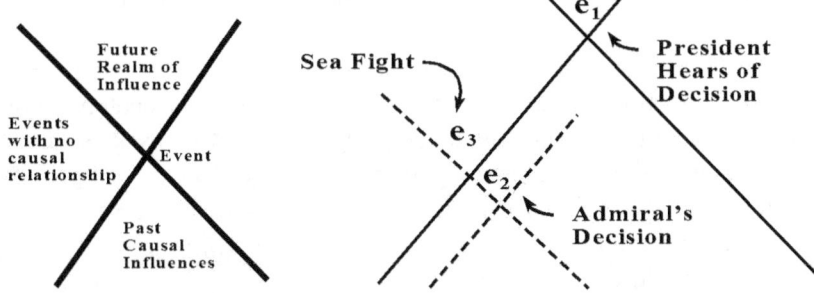

Figure 4.2. Left, the "light-cone" of an event. Right, the admiral's decision (e_2), the sea fight (e_3), and the president's learning of the decision (e_1).

On the right, three events (e's) are pictured. The dotted "X" is the light cone of e_2. Thus event e_3 lies within the top V of e_2's "X" – e_2 can causally influence e_3. Event e_1 also lies within the top V of e_2's "X". So e_2 can causally influence e_1. Event e_3 however is in the causal elsewhere of e_1. It cannot causally influence e_1. To make this a little more concrete: At event e_2 an admiral decides to engage the enemy ships, an event (the battle) which subsequently occurs at e_3. While the sea battle at e_3 is in the causal future of e_2, for e_1 it is in the area of events with no causal relationship. At e_1 the president learns of the admiral's decision (e_2). Pretending we are dealing with events moving at light speed here, we can easily imagine a movie scene where Roosevelt is consulting with his advisors, wondering what is going on in the Pacific theatre, and being told of Admiral Spruance's decision to engage at Midway. The entire scene takes place in this pregnant context of meaning, for we know simultaneously that the brave admiral and his fleet have already sailed into the fray. Thus, Rakić noted that it is clearly possible that e_3 (the sea battle) has already become, i.e., is *realized* with respect to e_1 (when the president hears about it). Though at e_1 the message about the battle has not yet been received, this does not imply that the sea battle has not occurred, i.e., is not realized. By this is meant that there is a very real, *ontological* relationship between e3 (the battle) and e1 (learning about the decision). The relativistic model of time is entirely insufficient to represent the actual fabric of time.

Rakić is content to term past, present and future not temporal but ontological relations, letting STR hold the ground as a theory of time. She notes however that her arguments could be construed as saying that STR is not a theory of time, allowing ontology and time to go together as is usual to common sense, but argues that in this case common sense must concede.

Holding Rakić's strained concession in place is undoubtedly STR's apparent success as an explanatory theory, all of which unfortunately revolves on asymmetric effects. But, the inadequacy of the Minkowski schema to represent time and its causal flows goes more deeply. When the flight of the fly, with his sixty-wing-beat distance invariant, was treated as an invariant causal flow, the beginning of the problem for the relativization of simultaneity was already introduced.

The Simultaneity of Flows

The intuition of a universal flow is partially preserved in relativity in the conservation of a "causal order." On analysis, we will find multiple causal orders or flows within this flow as Bergson noted or even, as Gibson insisted in the opening quote, where hero rushes to save the endangered heroine. The simultaneity of flows is integrally bound to causal order and to a global transformation wherein the motions of "objects" are transferences of state. Consider two football players running down each sideline of the field at precisely equal velocity. A physicist (O_1) at the fifty yard line notes the time against two synchronized clocks on each sideline as the players run by and ascertains that they have passed the same point simultaneously (Figure 4.3, e_1 and e_2). Of course a second physicist (O_2), thinking the first in motion and noting this observation says the first is in error, the events were not simultaneous. Yet the two football players continue on, converging on a football equidistant from both that they both kick simultaneously (e_3), kicking the ball twice as far as just one would have achieved. From the perspective of an instantaneous measurement, i.e., abstract time, their simultaneity is relativized. From the perspective of the two causal flows, the simultaneity of the flows is absolutely real. The second physicist cannot deny the effect of the simultaneous kick. One cannot simply relativize multiple causal flows.

It can be argued that e_1 and e_2 are not truly simultaneous just as O_2 states, that simultaneity is achieved only at the point-instant of the kick. But we could replace the football players equally well with a huge cue stick sweeping down the field towards a billiard ball. Positioned at each yard line are O_1's measurement clocks. If the cue's outside edges truly fail to pass the measurement clocks/ points simultaneously, it will hit the ball at a slant sending it off at an angle (Figure 4.4). In sliding the x_1, x_2 and t_2 axes upwards towards e_3, it can be seen that there will come a point as our very wide cue nears the ball at e_3, that e_3 will fall in the causal elsewhere of the light cones of each of the

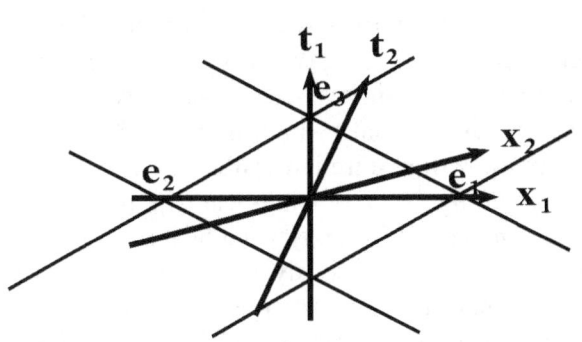

Figure 4.3. Two football players (e_1, e_2) converge on the ball (e_3).

edges (e_1, e_2). This implies that the two outer edges could not possibly be squared in time for a flush contact of the entire cue surface with the ball if they are as non-simultaneous as claimed by O_2. The global causal flow led by the cue's frontal surface is fragmenting under STR's treatment. Yet the cue strikes the ball precisely perpendicularly. Only one strand in this flow, one local flow, the causal order in STR invariant to both observers, is ultimately preserved. This is the chain of causal relations, <, the relation determining time-like and space-like events, defined upon a sequence of infinitely minute point-instants extending through the time line t_1 to e_3. Were we considering the fly, no matter how infinite the "points" we place on this line, or the in fact multiple lines comprising the fly, this will remain sixty wing beats – an indivisible movement or flow. A global flow, whether fly or cue stick or hero and heroine, cannot be an invariant to all observers in STR.[3]

We must ask what is the causal validity or efficacy of this one local point-instant flow? The breakup of simultaneity, as we have earlier seen, drives downwards in space-time to the most infinitesimal of point-instants. At this mathematical point, as earlier noted, there is neither time nor events. As such, without the possibility of even an event, it is impossible to say that there is anything causal whatsoever with respect to this point, or with respect to a "causal" chain of such

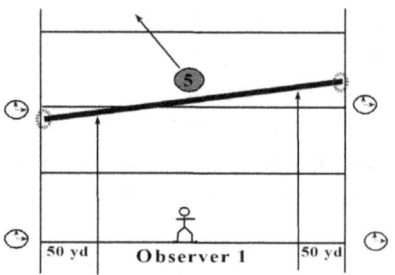

Figure 4.4. The flow of the very wide cue stick up the football field. If its edges are not simultaneously passing the yard lines, as Observer 2 believes, it must strike the ball at a slant.

points. The abstract space and abstract time that support the classical concept of causality offers again an infinite regress. If this chain is infinitely divisible – an infinite set of "point-events" – then between each point we must introduce a "causal relation," which is in effect to say a motion ad infinitum. Causality too will require indivisible extents. The fly, as a coherent biological system doing his sixty wing-beat trip, is precisely a global, indivisible flow. Were he taking his sixty wing beat trip to e_3, the tips of his wings will stop precisely simultaneously, O_2's measurements to the contrary notwithstanding. When it was insisted earlier that this sixty wing beat flow be treated as invariant to both X and Y, this weakness inherent in STR's treatment emerged.

In the above, I have not attempted a formal definition of a causal flow. I am leaving this at the intuitive level where, for example, a fly, as a complex system in motion, is comprised of multiple processes acting in concert, be this multiple muscle systems, neurons firing, or chemical flows. Such a system could be as large as, and larger than, a weather system such as a hurricane, or an evolving galaxy, or a collection of individuals all working together to play a symphony. The two football players with which we began were two seemingly isolable local flows. They could, however, have been two sailboats moving in unison before a vast pressure front. Or this could have been a vast magnetic flux sweeping the earth. The point is that we must ask if any such local flows, any more so than "objects" and their "motions," are truly isolable from the global time-flow of the universal field. Are they more than transferences of state within the global motion? This global transformation is the classical "flow of time" invariant to all observers.

STR and Consciousness

Two theorists, Hagan and Hirafuji, analyzed the concept of the "emergence" of consciousness in the context of relativity.[4] Emergence envisions consciousness arising (or being generated from) from the physical processes in the brain, analogous, it can be said, to the glow arising from the filament of a light bulb. That is, it is the same old "experience is generated by the brain" notion. Their analysis deals a critical blow to the emergence concept, but a deeper reading indicates that doubt is cast on STR's ability to support of any theory of consciousness.

Starting with what they term the *extrinsic* definitional problem, they argue that any emergent state of consciousness (experience) must be frame invariant to satisfy the requirement that the conscious state be invariant to another observer in motion. What this means, let me immediately interject, is simply that the simultaneity of all the musicians playing a symphony that one is watching better be maintained! Hagan and Hirafuji aver that keeping the emergent property frame invariant might be achieved, but choose not to explore the difficulty, moving on to yet another extrinsic problem (what they term a "boundary" problem). In fact, it cannot be achieved. Our experience, we have seen, is marked by the characteristic of simultaneity of flows – the multiple melody lines within a single flow of a symphony, multiple musicians playing on the symphony stage, multiple women cooking in the kitchen, etc.

From the standard view of relativity, from which Hagan and Hirafuji write, the simultaneity of any of the above systems (read *experiences* as well) should indeed breakup, simultaneity becoming succession, and succession becoming simultaneity. Recall the three points, A', B', and C' of Figure 4.1, and the break up of simultaneity at the most infinitesimal interval. We asked if this can possibly be true of the flow of time? In the more obvious causal context of causal flows, e.g., our two football players, we saw that this cannot be true. One cannot simply relativize multiple causal flows.

Yet this is precisely what relativity would do. Each of the experiences mentioned earlier, with their simultaneous flows, would begin to breakup relative to the motion, for example, of observer Y. This is why the "emergent" consciousness or emergent "property," as Hagan and Hirafuji mention, would have such difficulty remaining frame invariant. More correctly, this is why the invariance is impossible. The experience would inevitably be distorted relative to the frame. But as I asked earlier, can we seriously believe this "breakup" of succession and simultaneity is possible, i.e., *that it has any ontological status!*? Do we believe the symphony would become jumbled, the musicians playing out of time, the conversations at the table scrambled, the cooking women putting ingredients in the cake one after the other rather than together, etc.?

One could question the relevance of the frame invariance requirement. So what, if from Y's point of view, my consciousness is distorted? It is my consciousness and it is perfectly OK, the symphony is fine, the ladies' conversation is fine. But this is the problem: if the theory (STR) is taken to indicate that this distortion would indeed be so from Y's perspective, i.e., it has ontological status, despite the intuitive oddity of the claim, we must ask what good is the theory? Hagen and Hirafuji are not only demonstrating the difficulties with a theory of "emergence" in the context of current physical theory, but also the difficulties for relativity of supporting any model of consciousness.

Let us move to the *intrinsic* definitional problem. Hagan and Hirafuji show that an intrinsic definition, while not requiring simultaneity, will always be incompatible with locality constraints. The difficulty here stems from the transmission speeds of the brain or, simply the very need or constraint for finite transmission. Under these constraints, the brain could not support a global state underlying an emergent property. The global state cannot inform the local dynamics of the boundary necessary to establish the physical extent of the emergent unit. But in essence here, I note, we have come back to the need for simultaneity, for this is an essential feature of any emergent property of consciousness or perception of which we can conceive.

Stein, as we saw, attempted to explain ongoing misconceptions of relativity, as he saw them, in terms of our continued naïve belief in the perception of simultaneous events – an illusion based on the high velocity of light.[5] Thus, he argued in essence, the naïve or intuitive simultaneity that perception provides is founded upon the "fleeting motions" of "masses of elements" in the brain, all subject to the limitation of communication via the velocity of light and implying, therefore, that at a small enough scale of time, perceptive simultaneity would break down. Stein is assuming a model of the processes in the brain underlying perception. But it is precisely this "fleeting motion" of masses of "elements" that Hagan and Hirafuji demonstrate is subject to locality constraints and, in being so subject, cannot support the simultaneity inherent in conscious states or perception, at least not from an "emergence" standpoint. If we only require a classical dynamics within the brain, under the locality constraint, to support a specifying reconstructive wave, as per Bergson's model, we escape the emergence difficulty, but this framework, with its non-differentiable time and simultaneity of flows, leaves relativity behind.

Chapter IV: End Notes and References

1. Bergson,H. (1923), *Duration and Simultaneity*, p. 103.
2. Rakić, N. (1997). Past, present, future, and special relativity. *British Journal for the Philosophy of Science, 48*, 257-280.
3. A comment on concepts expressed by Myrvold is appropriate here. Myrvold considers the relation eRe' (where R = "realized with respect to") in the context of *extended* objects. This requires taking a spacelike slice – in effect an instantaneous stage along some foliation of the object's history. Failure to do this results, he notes, in paradoxes like the "pole and barn," where, with the barn at rest and the pole in high velocity motion through the barn, there is a period where the pole just fits inside the barn, and conversely, with the pole at rest, and the barn in motion, there is no such "pole-inside" state. This conflict is resolved, he argues, "by remembering that the states of the extended system of which one account speaks are states along spacelike slices of the system different from those of which the other account speaks" (p. 478).

    This is a not a justifiable modification of STR. The reference system of Figure 7.8 would be treated as a set of points, α. Another set, β, would be definite or realized with respect to α if in α's causal past. Though seemingly applying to the cue stick example, we could not extend the system indefinitely, or it would extend across the entire universe, providing a plane of simultaneity. But, given Myrvold, what prevents this move? My earlier analysis relative to Figure 7.8 shows that the simultaneity of α begins to break up at the most minute interval relative to an observer in motion. But, there is a simpler reason why Myrvold is not a resolution. If the length contraction of the pole is being taken as a real effect in this paradox, the (very testable) implication is that we could actually trap the pole inside the barn, different spacelike slices or not. Such a real result (captured pole) is as much a contradiction as the twin paradox. If it is not considered a real (possible) effect, this is due to giving the reciprocity of reference systems its appropriate status, which is to say there is no ontological status to the relativistic contraction, and no "paradox" in the first place. Myrvold dismisses the paradox, considering it an example of misunderstanding, yet it is no more a misunderstanding than the twin-paradox where the "time-change" should have equally as little ontological status.

    Myrvold, W. (2003). Relativistic quantum becoming. *British Journal of the Philosophy of Science, 54*, 475-500.
4. Hagan, S., & Hirafuji, M. (2001). Constraints on an emergent formulation of conscious mental states. *Journal of Consciousness Studies, 8*, 109-121.
5. Stein, H., op.cit, 1991.

CHAPTER V

Conclusion

It is the lasting achievement of the Bergsonian metaphysic
that it reversed the ontological relation assumed between
being and time.

— Ernst Cassirer, *The Philosophy of Symbolic Forms*[1]

The Singular Time of Consciousness *and* Physics

There have been other examinations of STR, of both its explanatory status in physics and as a theory of time. Bergson was perhaps the earliest. Deleuze would reprise Bergson's general argument on time with respect to relativity.[2] Dingle would make interesting critiques, particularly on the invariance of light.[3] Brillouin would give a non-relativistic explanation of the retardation of atomic clocks (and of the red shift).[4] Earman would note that there has yet to be a relational, let alone a relativistic explanation of Newton's humble bucket.[5] Nordenson would argue that Einstein's rejection of the classical flow of time, whether beyond "proximity" or anywhere even beyond the mathematical point, must surely undermine any meaning to his new procedure for clock synchronization.[6] Rakić, in proving the logical inadequacies of the Minkowski metric, is reduced to declaring Special Relativity to be not an ontological theory, but concedes it a status as a "temporal" theory. Whatever meaning this concession might have, a theory with no ontological status is of little use; it is certainly not relevant to a science of perception or a theory of consciousness.

For anyone accepting and using relativity in their thought on space and time, or trying to integrate it with quantum theory, you must ask, *just what theory are you using*? If it is the theory that is supposedly supporting relativistic effects as ontological or real, it is an utter mess of contradictions, and begun so by Einstein himself. This is not to mention the problems then generated by the implied reality of relativized simultaneity. If it is the true *Invariantentheorie* with its self-consistent use of invariance under transformation, it is useless as a theory to explain such effects as ontological, yet we know these exist. I cannot comprehend how one can fail to ask this *What theory*? question. Further, it is a theory that cannot support the obvious reality of simultaneous causal flows and it is a theory that is simply an extension of an outmoded metaphysic that will never support consciousness, perception or the quality of the perceived world. These are absolute requirements for a theory of time.

STR, with its confused interpretation and its reflection of the classic, spatial metaphysic with its degenerate view of "time" is an impediment to both physics and psychology. It is a mist that has arisen over the fields of theory, a mist that has beclouded all theory. Physics has struggled to both reconcile STR/GTR with quantum theory (aggravated by the awareness of quantum theory's non-locality) and simultaneously to understand and perhaps incorporate the role of consciousness in quantum theory. The mist has crept into the theory of consciousness.

Conclusion

The theory of time is precisely the ground where psychology, the theory of consciousness and physics meet. In truth, with Bergson's vision of time – with its non-differentiable flow, with its irreversibility derived from the fact that each "instant" reflects the entire preceding series, with its primary memory or true continuity wherein there are no mutually external "instants," where the motions of "objects" are transferences of state within a global time-evolution of the material field implying therefore an inherent non-locality – one sees that Einstein's two times, "a psychological time different from that of the physicist," are in reality one.

Chapter V: End Notes and References

1. Cassirer, E. (1929/1957). *The Philosophy of Symbolic Forms, Vol. 3: The Phenomenology of Knowledge.* New Haven: Yale University Press, p. 184.
2. Deleuze, G. (1966/1991). *Bergsonism.* (Translated by H. Tomlinson and B. Habberjam) New York: Zone Books.
3. Dingle, H. (1972). *Science at the Crossroads.* London: Martin Brian & O'Keeffe.
4. Brillouin, L. (1970). *Relativity Reexamined.* New York: Academic Press, pp. 77-85.
5. Earman, J. (1989). *World Enough and Space-time.* Cambridge: MIT Press.
   Newton noted how, when a bucket filled with water is spun rapidly, the water is flung (by the centrifugal force) in a standing wave pattern against the sides of the bucket. How, he asked, can this motion/force of the water be relativized with respect to the bucket? It is an absolute motion or effect.
6. Nordenson, H. (1969). *Relativity, Time and Reality.* London: George Allen and Unwin.

www.ingramcontent.com/pod-product-compliance
Lightning Source LLC
Chambersburg PA
CBHW071621170526
45166CB00003B/1139